Analysis of Watersheds Monitored by the U.S. Geological Survey Streamflow-gaging Station Network in the Upper Colorado River Basin

By Terry A. Kenney, Susan G. Buto, and David D. Susong

Scientific Investigations Report 2011–5081

U.S. Department of the Interior
U.S. Geological Survey

U.S. Department of the Interior
KEN SALAZAR, Secretary

U.S. Geological Survey
Marcia K. McNutt, Director

U.S. Geological Survey, Reston, Virginia: 2011

For more information on the USGS—the Federal source for science about the Earth, its natural and living resources, natural hazards, and the environment, visit http://www.usgs.gov or call 1–888–ASK–USGS.

For an overview of USGS information products, including maps, imagery, and publications, visit http://www.usgs.gov/pubprod

To order this and other USGS information products, visit http://store.usgs.gov

Suggested citation:
Kenney, T.A., Buto, S.G., and Susong, D.D., 2011, Analysis of watersheds monitored by the U.S. Geological Survey streamflow-gaging station network in the Upper Colorado River Basin: U.S. Geological Survey Scientific Investigations Report 2011–5081, 47 p.

Contents

Figures

Tables

Conversion Factors

Inch/Pound to SI

Multiply	By	To obtain
Length		
inch (in.)	2.54	centimeter (cm)
foot (ft)	0.3048	meter (m)
mile (mi)	1.609	kilometer (km)
Area		
square mile (mi^2)	2.590	square kilometer (km^2)
Volume		
acre-foot (acre ft)	1,233	cubic meter (m^3)

Vertical coordinate information is referenced to the North American Vertical Datum of 1988 (NAVD 88).

Horizontal coordinate information is referenced to the North American Datum of 1983 (NAD 83).

Analysis of Watersheds Monitored by the U.S. Geological Survey Streamflow-gaging Station Network in the Upper Colorado River Basin

By Terry A. Kenney, Susan G. Buto, and David D. Susong

Abstract

The U.S. Geological Survey (USGS) has operated streamflow-gaging stations in 1,053 watersheds in the Upper Colorado River Basin (UCRB) since 1894. Currently, 223 of these streamgages are active. This report presents selected watershed characteristics for 10,338 watersheds in the UCRB. These watersheds are compared to the watersheds upstream of USGS streamgages to assess how well the USGS streamgage network represents the physical characteristics of the watersheds in the entire basin. To conduct this assessment, 17 watershed characteristics, including physiographic parameters, land cover types, lithology, and parameters that describe anthropogenic influence, were computed for each of the gaging station drainage basins. The set of 10,338 watersheds in the UCRB was constructed from a previously developed stream-reach network, and the same 17 basin characteristics were computed for each watershed to facilitate comparisons.

The USGS streamgage watersheds and the UCRB watersheds were split into those that are currently unaffected by upstream reservoir regulation and those currently affected by upstream reservoir regulation. In general, for unregulated watersheds, the streamgage network represents the range of most basin characteristics in the watersheds of the UCRB. However, the active streamgage network for unregulated watersheds is generally lacking in representation of most basin characteristics compared with watersheds in the UCRB. At regulated locations, the streamgage network including the active network, generally represents the range of most basin characteristics well.

Introduction

The U.S. Geological Survey (USGS) began gaging streamflow in the Upper Colorado River Basin (UCRB) with the installation of the Green River at Green River, Utah, streamgage in 1894. Since that time, the USGS has operated streamgages at more than 1,050 sites in the UCRB. The purposes for these gaging stations vary and have changed with the needs for differing water information over the last century.

In the late 19th and early 20th centuries, gaging stations in the UCRB were established to quantify the amount of water in the basin as a result of the Sundry Civil Appropriations Act of 1888. The act mandated that the USGS, under the direction of the Secretary of the Interior, investigate the extent to which the arid region of the United States could be redeemed by irrigation and for selecting sites for reservoirs and other hydraulic works (Corbett and others, 1943). Following the signing of the Colorado Compact in 1922, gaging stations were established to assist the Upper Basin States of Colorado, Utah, Wyoming, New Mexico, and Arizona, with accounting for their allocation of Colorado River water. As water development projects, specifically reservoirs, were conceived, streamgages were established to aid in both assessing the utility of sites and ultimately, in the design of the structures. As the population of the region grew both within and bordering the UCRB, for example, in the Colorado Front Range, the need to develop municipal water supplies expanded the USGS streamgage network in the UCRB. The number of gaging stations also increased to assist in managing and distributing water, specifically for agricultural use and water rights management. Infrastructure improvements, such as highways and roads, in the middle of the 20th century also led to the establishment of streamgages to provide data for the design and construction of bridges. Streamflow data in the latter part of the 20th century and into the 21st century are used for understanding the relation between streamflow and water quality, streamflow and climate, and streamflow and ecology. Over time there has been an evolution in water resource issues, and the USGS streamgage network has evolved and changed with the issues. With the changing needs for streamflow data, existing gages are discontinued, new gages are established, and older gages, in some cases, are reactivated. In addition to the scientific and engineering needs for streamgages, socioeconomic factors have influenced the growth and decline of the USGS streamgage network in the UCRB.

Purpose and Scope

The purpose of this report is to present the results of an examination of the USGS streamgage network, both past and present, in the UCRB in terms of the physical watershed characteristics of the UCRB. The watersheds monitored by the current network of active streamgages and watersheds previously monitored by inactive USGS streamgages are compared with watersheds in the UCRB in terms of drainage basin size and selected watershed characteristics. The UCRB is subdivided into a set of 10,338 watersheds, which are compared to the subset of watersheds instrumented with USGS streamgages. The objective of this comparison is to assess how well the watersheds instrumented with USGS streamgages represent the varied landscapes in all UCRB watersheds.

Description of Study Area

The Colorado River Basin, which drains portions of seven states, is the largest river basin in the southwestern United States. The UCRB, for purposes of this assessment, is defined as the drainage basin upstream of USGS streamflow-gaging station 09380000, Colorado River at Lees Ferry, Arizona. The UCRB has a contributing drainage area of about 108,000 mi^2 and includes parts of Wyoming, Colorado, Utah, New Mexico, and Arizona (fig. 1). The UCRB drains a large portion of the Rocky Mountains west of the Continental Divide, from the Wind River Mountains in Wyoming south to the San Juan Mountains in Colorado. Major drainages in the UCRB include those of the Colorado, Green, and San Juan Rivers.

U.S. Geological Survey Streamgages in the Upper Colorado River Basin

The earliest computed daily mean discharge value in the UCRB is for October 1, 1894, from USGS streamflow-gaging station 09315000, Green River at Green River, Utah. As of 2010, there were 1,067 USGS streamgages in the UCRB, with at least 365 daily mean discharge values contained in the USGS National Water Information System (NWIS) database (fig. 2; table 1; Appendix A). This compilation of streamgages includes both active and inactive gages, as well as 14 inactive gages whose locations are currently inundated by reservoirs. Active gages are those in operation by the USGS in 2010. Inactive gages are gages that have been historically operated by the USGS and have a minimum of 365 days of daily mean discharge values. These inactive gages have varying periods of record (Appendix A). Of the 1,053 streamgages in the UCRB, 223 were active in 2010 (U.S. Geological Survey National Water Information System; http://nwis.waterdata. usgs.gov). Thus, for the purposes of this report, the USGS streamgage network is referenced in three ways: the entire USGS streamgage network, which includes 1,053 inactive and active gages; the inactive USGS streamgage network, which includes 830 inactive gages; and the active USGS streamgage network, which includes the 223 gages in operation in 2010. Streamgages located on unnatural watercourses, such as canals, are not included in this analysis.

There are 80 water-storage reservoirs greater than 5,000 acre-ft in the UCRB (Ruddy and Hitt, 1990). Agricultural activities and the human population in the UCRB are dependent upon these reservoirs. Water storage in these reservoirs substantially alters the natural streamflow patterns of the streams they impound as well as those downstream. Of the 1,053 streamgages in the UCRB, 226 gages are located downstream of reservoirs with capacities of at least 5,000 acre-ft (Ruddy and Hitt, 1990). To distinguish between regulated and unregulated streamflow in this investigation, streamgages and watersheds in the UCRB were divided into two groups: 1) unregulated, which are gages and watersheds that are unaffected by upstream reservoir regulation, and 2) regulated, which are gages and watersheds that were affected by reservoir regulation as of 1990 (Ruddy and Hitt, 1990). Reservoir regulation, as defined for this study, pertains to gages that are located downstream of at least 5,000 acre-ft of reservoir capacity. Other human influenced water development, such as diversion canals and transbasin diversions, was not considered in the regulated/unregulated classification of watersheds. Because of the availability of data on reservoirs in the UCRB and because, in general, unless the watershed is very small, a reservoir smaller than 5,000 acre-ft does not have much affect on downstream streamflow, 5,000 acre-ft was selected as the minimum amount of reservoir capacity to be considered a regulated watershed in this report.

Drainage Basin Areas for Watersheds with and without U.S. Geological Survey Streamgages in the Upper Colorado River Basin

Drainage basin area is a fundamental landscape characteristic that is used when classifying or studying streams and streamflow. In this study, drainage area is used as the first-order watershed descriptor by which other landscape characteristics are presented. All USGS streamgage locations have the contributing drainage area computed; drainage areas for the set of UCRB watersheds were also computed.

Drainage Areas for U.S. Geological Survey Streamgages Unaffected by Reservoir Regulation

Drainage areas for the 827 unregulated streamgage locations in the UCRB range from 0.03 to 4,040 mi^2. In 2010, 125 of these gages were active. Drainage areas for active gage locations range from 2.27 to 4,040 mi^2. The number of unregulated streamgages by drainage area, grouped in 200 mi^2 increments, is shown in figure 3. There are 112 currently active

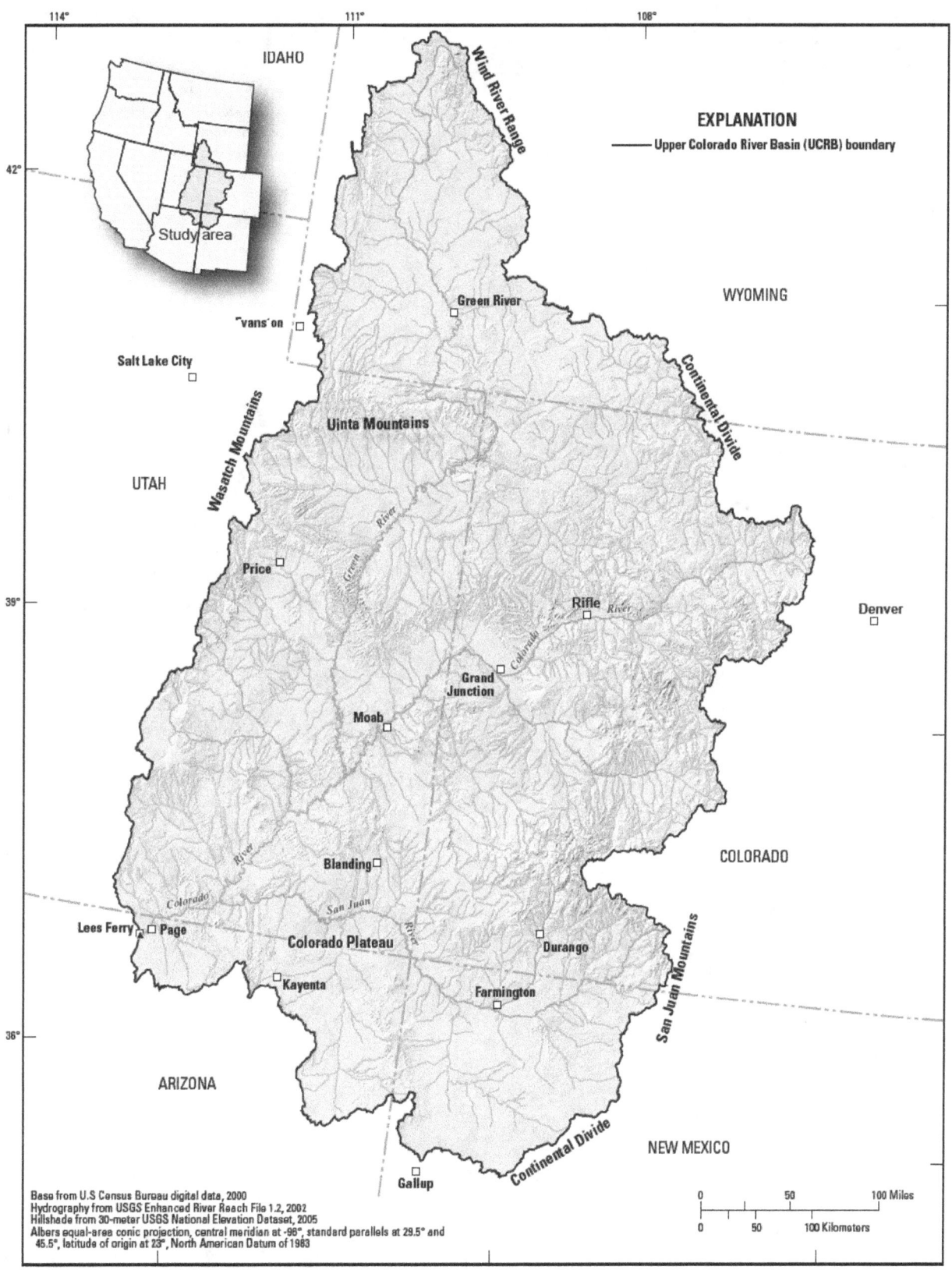

Figure 1.   Upper Colorado River Basin study area.

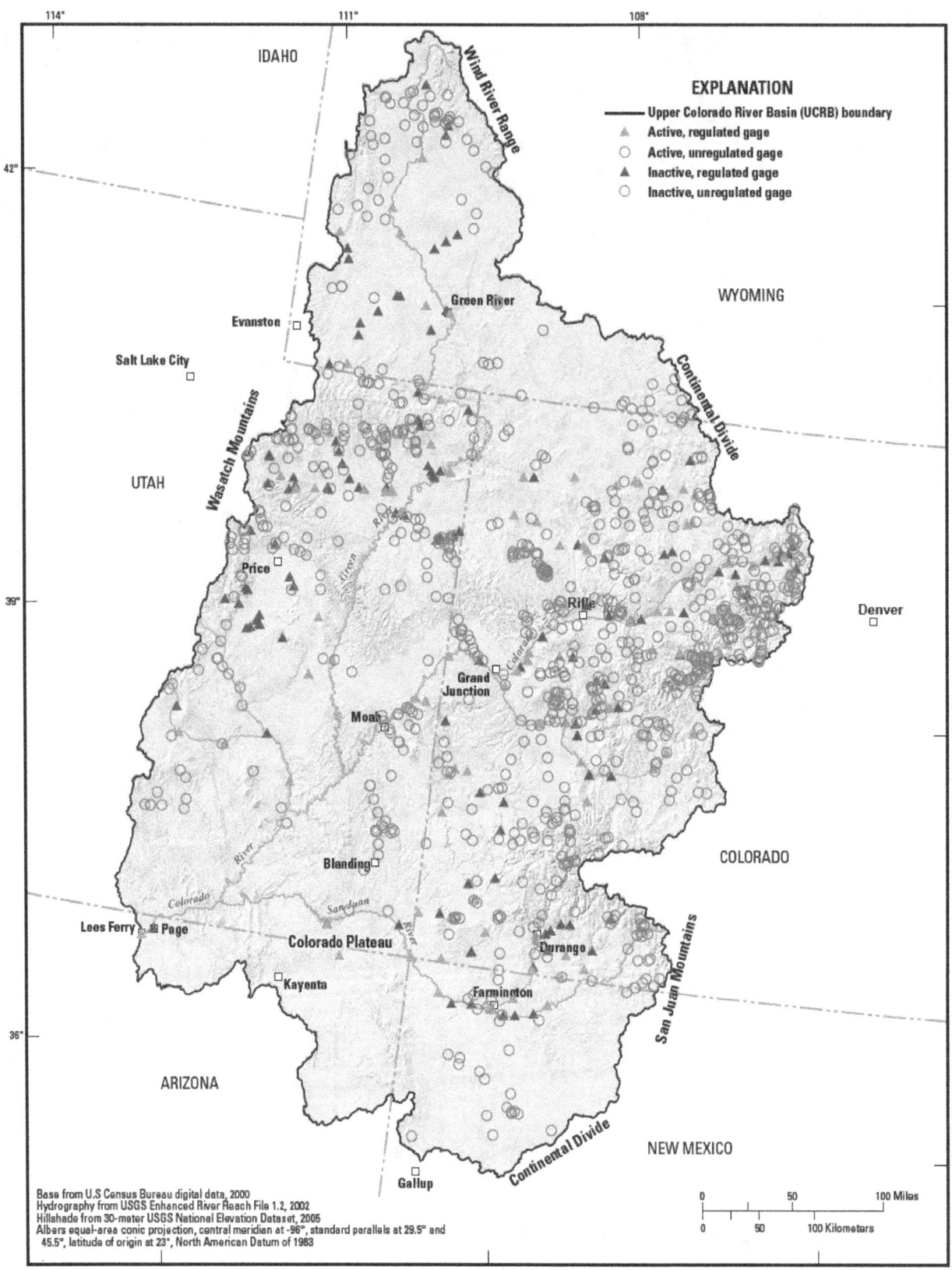

Figure 2.   U.S. Geological Survey streamflow-gaging station network for the Upper Colorado River Basin.

Table 1. Number of unregulated and regulated U.S. Geological Survey streamflow-gaging stations and Upper Colorado River Basin watersheds examined.

Site	Total number	Number where outlet is not inundated by reservoir	Number that are unregulated	Number that are regulated
U.S. Geological Survey streamgages	1,067	1,053	827	226
Active	223	223	112	98
Inactive	844	830	715	128
Upper Colorado River Basin watersheds	10,679	10,338	8,985	1,353

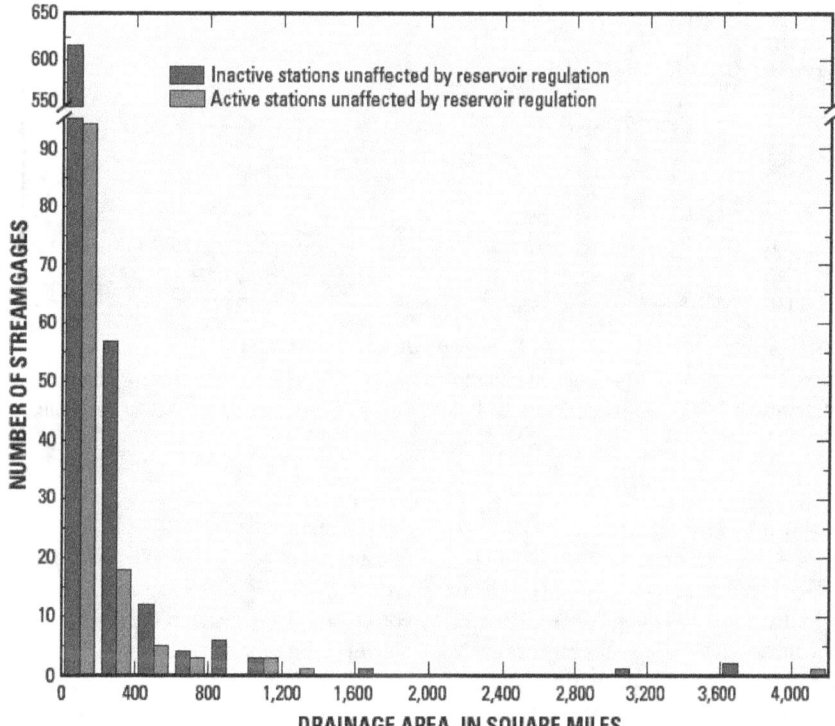

Figure 3. Number of streamgages unaffected by reservoirs (unregulated) with normal capacities of 5,000 acre-feet or more by drainage areas.

stations out of 785 stations that have drainage areas ranging from 0.03 to 400 mi². The average period of record for the 785 gages is 17 years, whereas the average period of record for the 112 active gages is 42 years. The number of streamgages having drainage areas from 0.03 to 400 mi², grouped in 25 mi² increments, is shown in figure 4. There are only 25 currently active streamgages out of 332 that have drainage areas less than 25 mi² in the UCRB. The 332 streamgage locations having drainage areas less than 25 mi² have an average period of record of about 13 years, whereas the 25 active stations have an average period of record of about 32 years. Finally, there are 88 gages, 4 of which are active, in drainage areas less than 5 mi².

Drainage Areas for U.S. Geological Survey Streamgages Affected by Reservoir Regulation

Of the 1,053 USGS streamgages in the UCRB, 226 are downstream of water-storage reservoirs with a normal capacity of at least 5,000 acre ft (Ruddy and Hitt, 1990). The period of record for many of these gages begins prior to any upstream water development. Thus, data from these gages potentially represent a mix of unregulated and regulated streamflow conditions, but for the purposes of this investigation, all of these 226 streamgages are considered affected by reservoir regulation. Drainage areas for the gages affected by upstream reservoir regulation range from 8.2 to 108,000 mi². Of the 226 gages affected by regulation, 98 active gages have

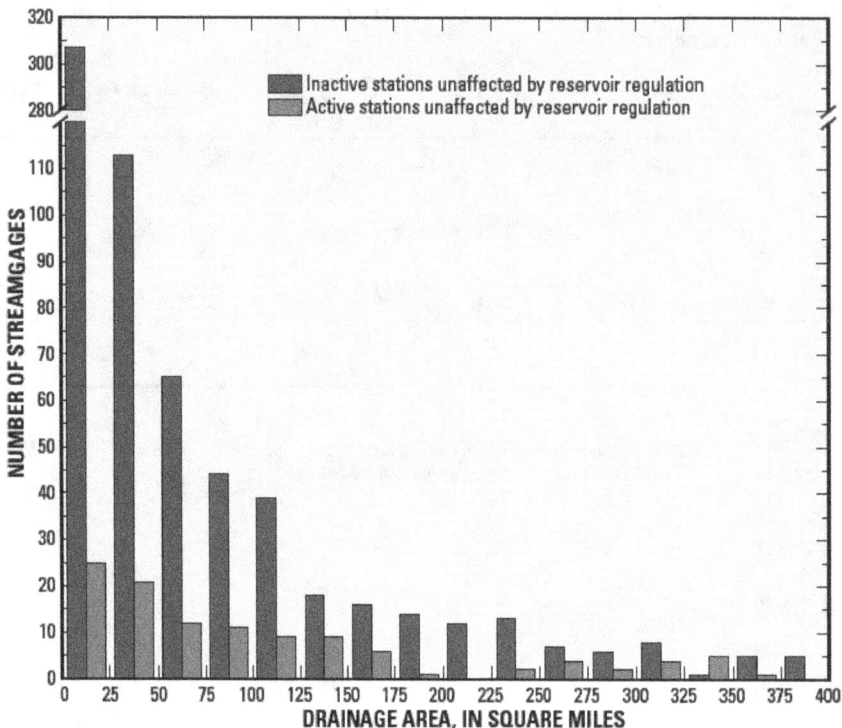

Figure 4.   Number of streamgages unaffected by reservoirs (unregulated) with normal capacities of 5,000 acre-feet or more in drainage areas between 0.03 and 400 square miles.

drainage areas ranging from 36.7 to 108,000 mi^2. The number of regulated streamgages by drainage area, grouped in 5,000 mi^2 increments, is shown in figure 5. A total of 15 gages, 10 of which are active, have drainage areas greater than 10,000 mi^2. Only two gages, one of which is active, have drainage areas greater than 50,000 mi^2. The number of regulated streamgages with drainage areas less than 10,000 mi^2, grouped in 1,000 mi^2 increments, is shown in figure 6.

Watersheds in the Upper Colorado River Basin

To assess the representation of the watersheds in the UCRB by the USGS streamgage network, the watersheds with streamgages need to be compared with a dataset that represents all of the watersheds within the UCRB. For this assessment, a dataset of the watersheds in the UCRB was created from a synthetic stream-reach network consisting of 10,679 incremental catchments that were developed for a dissolved-solids surface-water quality model of the UCRB (Kenney and others, 2009). The stream-reach network consists of stream reaches and associated incremental catchments. These incremental catchments are bound upstream and downstream by other catchments and, therefore, do not represent entire watersheds unless they are the upstream-most, or headwater, catchment (fig. 7). The drainage areas of the incremental catchments of the synthetic stream-reach network ranged in size from 0.5 mi^2 to 78 mi^2, with an average area of about 10

mi^2 (Kenney and others, 2009). This network of unique stream reaches is linked by to- and from- nodes and a hydrologic sequence number that allows for downstream accumulation of constituent mass (used in the water-quality model) or, for the benefit of this analysis, provides a means to sum the upstream watershed characteristics of the incremental catchments to create a dataset of watersheds. By taking advantage of the linkages and sequencing of this network, a set of 10,679 watersheds of the UCRB was created for this assessment (fig. 8), and the areas of each of these watersheds were computed. The drainage areas of the resulting watershed dataset range in size from 0.5 mi to 108,000 mi^2. As would be expected, all of the watersheds are nested in watersheds that are located in the downstream direction.

Once the UCRB watershed dataset was constructed, it was compared to locations of reservoirs in the UCRB. Using the 24,000-scale National Hydrography Dataset (NHD 24k; U.S. Geological Survey, 1999), 341 of the 10,679 watersheds were found to have the outlet location inundated by one of the many reservoirs in the UCRB. These watersheds were removed from the analysis, leaving 10,338 watersheds to compare with the 1,053 watersheds represented by the USGS streamgage network in the UCRB. The practical lower limit of drainage basin area for the UCRB watershed dataset is about 5 mi^2. This is a function of the reach network and catchments that were used to generate the UCRB watershed dataset. Therefore, most comparisons of the UCRB streamgage net-

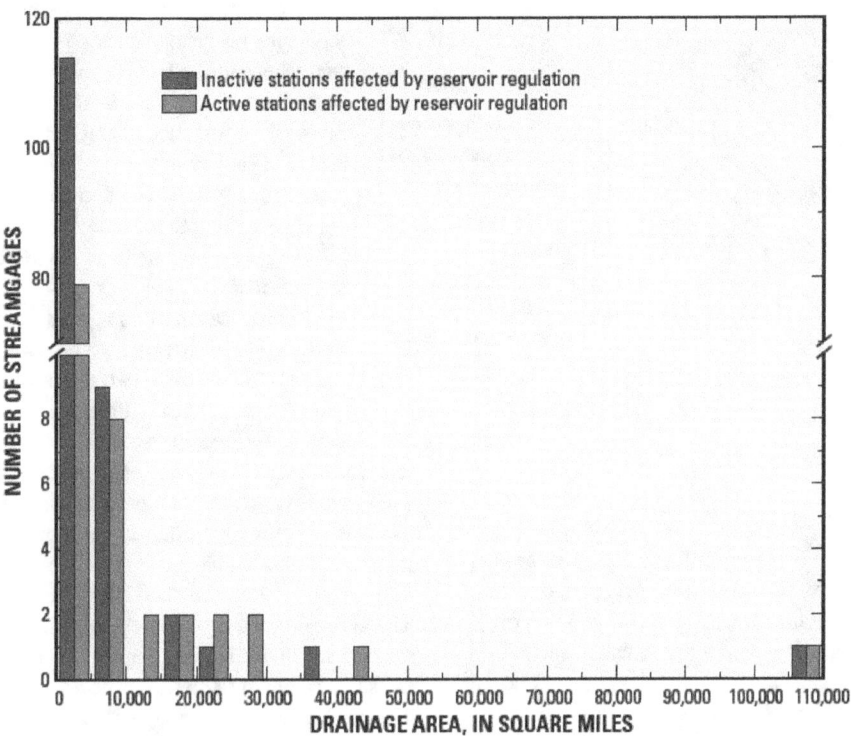

Figure 5.   Number of streamgages affected by reservoir storage (regulated) of 5,000 acre feet or more by drainage areas.

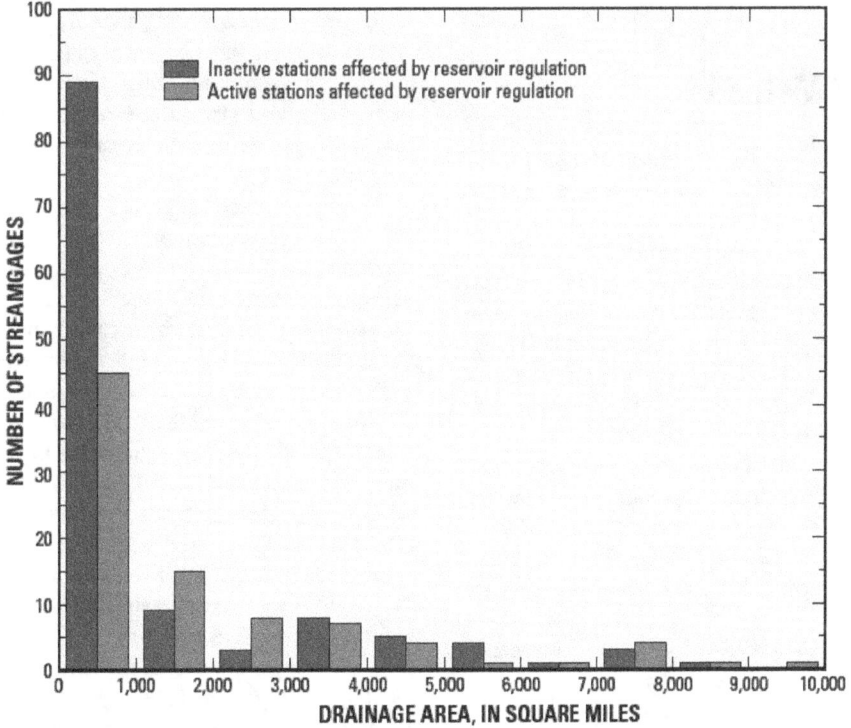

Figure 6.   Number of streamgages affected by reservoir storage (regulated) of 5,000 acre feet or more in drainage areas between 36.7 and 10,000 square miles.

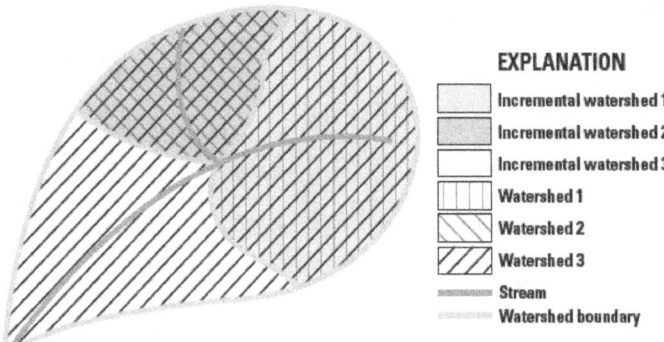

EXPLANATION

Incremental watershed 1

Incremental watershed 2

Incremental watershed 3

Watershed 1

Watershed 2

Watershed 3

Stream

Watershed boundary

Figure 7.   Diagram showing incremental catchments composing a watershed.

work with the watershed dataset will focus on basins greater than 5 mi^2. The number of unregulated and regulated watersheds in the UCRB by drainage area, grouped in 5,000 mi^2 increments, is shown in figure 9. All unregulated watersheds in the UCRB have drainage areas that are less than 5,000 mi^2, which is similar to the largest drainage area for an unregulated USGS streamgage of 4,040 mi^2. About 900 regulated UCRB watersheds have drainage areas that are less than 5,000 mi^2. The number of watersheds by drainage area that are less than 5,000 mi^2, grouped in 250 mi^2 increments, is shown in figure 10.

Approach and Methods

Characteristics of watersheds instrumented with USGS streamgages are compared with characteristics of watersheds in the UCRB watershed dataset to assess how well the USGS streamgage network represents the varied landscapes in the UCRB. These comparisons are presented graphically in scatter plots with corresponding bar graphs for each watershed characteristic for both regulated and unregulated watersheds. The scatter plots are constructed with drainage area on the y-axis and the watershed characteristic, such as mean basin elevation or percent evergreen forest cover, on the x-axis. Computed characteristic values for watersheds with active USGS gages are represented by solid red circles, and computed values for watersheds with inactive USGS gages are represented by solid blue circles. Computed characteristic values for the UCRB watersheds that do not have USGS streamgages are shown as light gray circles.

This graphical presentation provides a visual representation of how well the USGS streamgage network, both current and historical, captures the varied landscapes of the watersheds in the UCRB. Gaps and (or) bias in representation of specific watershed characteristics and drainage areas by the USGS streamgage network can be easily identified using the graphs. For example, if the characteristic values for watersheds

instrumented with USGS gages are evenly distributed across the range of values and drainage areas for all watersheds in the UCRB, this indicates that watersheds instrumented with USGS streamgages are representative of the watershed characteristic. Conversely, if there are gaps or the watersheds with USGS gages are biased for a watershed characteristic, it indicates that the USGS streamgage network is not repesenting the full range of watersheds in the UCRB for that watershed characteristic.

To provide a more quantitative measure of how well the USGS streamgage network represents the watersheds in the UCRB, the ratio of the number of USGS streamgages to the number of watersheds for each watershed characteristic was computed at various intervals along the range of each watershed characteristic. This ratio is presented as a percentage and is represented with bar graphs at the top of the scatter plots. The bar graphs share the same x-axis as the scatter plots and the bars correspond to intervals of the watershed characteristic being examined. The green bar is the percentage of UCRB watersheds instrumented with USGS gages (entire USGS network) that share the same range of values of a specific watershed characteristic. The red bar is the percentage of UCRB watersheds having active USGS gages that share the same range of values of a specific watershed characteristic. Drainage areas below 5 mi^2 were not included in the calculations for the bar graphs. The watershed characteristic values shown for a gaging station (red and blue dots) may not exactly correspond to, or overlap, the watershed characteristic for a watershed in the UCRB because streamgages may not be located in the exact location on the stream as the watershed outlet. Because of this, the bar graphs, particularly where the number of UCRB watersheds is small, may indicate a larger representation of the watershed characteristic by the gage network.

Scatter plots and bar graphs for each watershed characteristic were developed for both regulated and unregulated watersheds and are discussed in following sections of the report. The discussion for each of the watershed characteristics is limited to brief general statements about how well the USGS streamgage network represents watershed characteristics in the UCRB as shown by the scatter plots and bar graphs.

To assess how well the USGS streamgage network represents a specific watershed characteristic, a comparison of the number of watersheds with USGS streamgages to the total number of watersheds that share the same values of a specific watershed characteristic is used. By treating each of the watersheds in the UCRB as an independent watershed with the potential for locating a streamgage, it can be determined that the entire USGS streamgage network measures streamflow in about 9.2 percent of the unregulated watersheds in the UCRB (827 gages and 8,985 watersheds). The active USGS streamgage network measures streamflow in about 1.4 percent of the unregulated watersheds in the UCRB (125 gages and 8,985 watersheds). For a specific watershed characteristic, if the ratio, presented as a percentage, of unregulated USGS

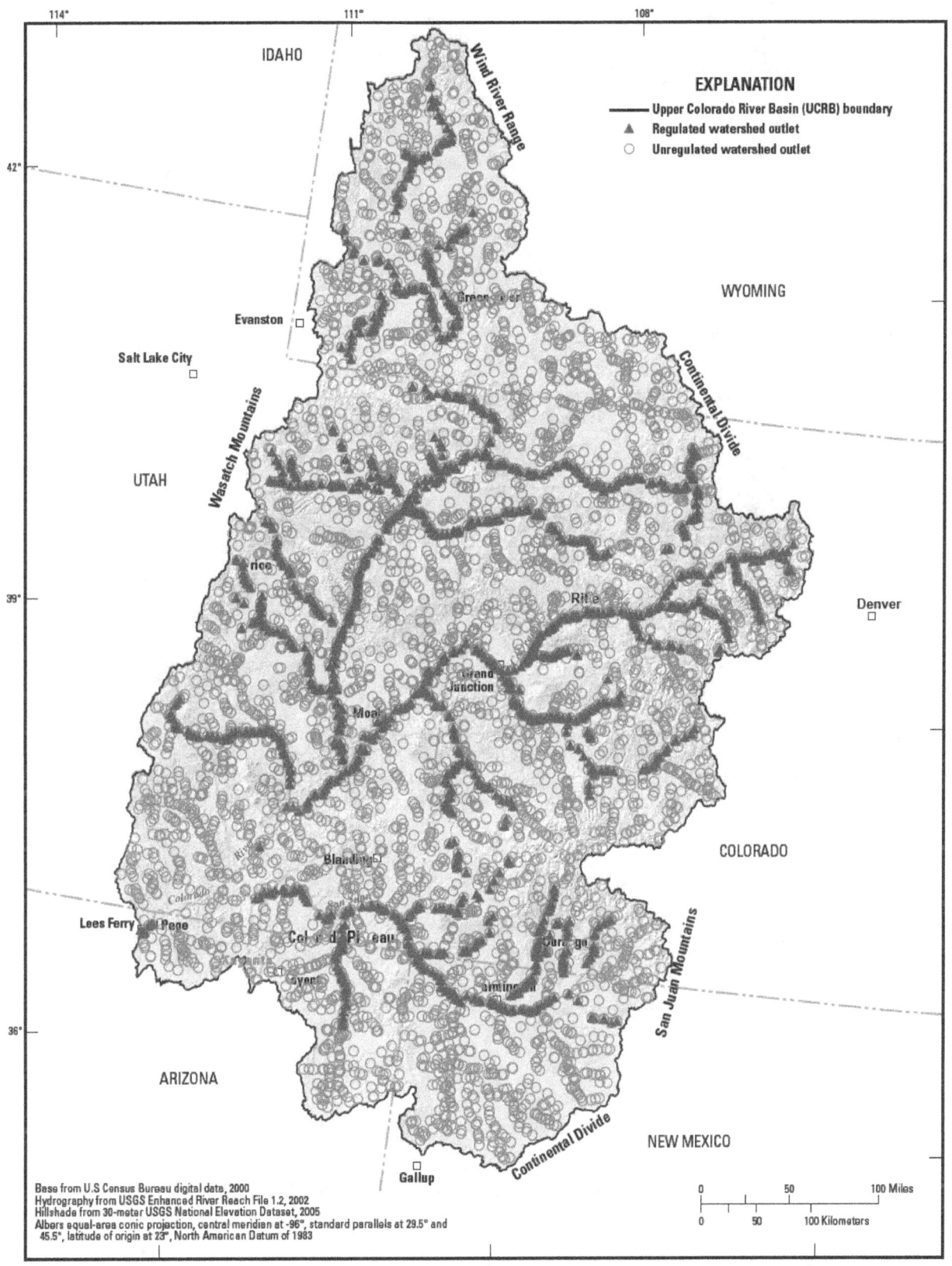

Figure 8.   Outlets of the watersheds of the Upper Colorado River Basin.

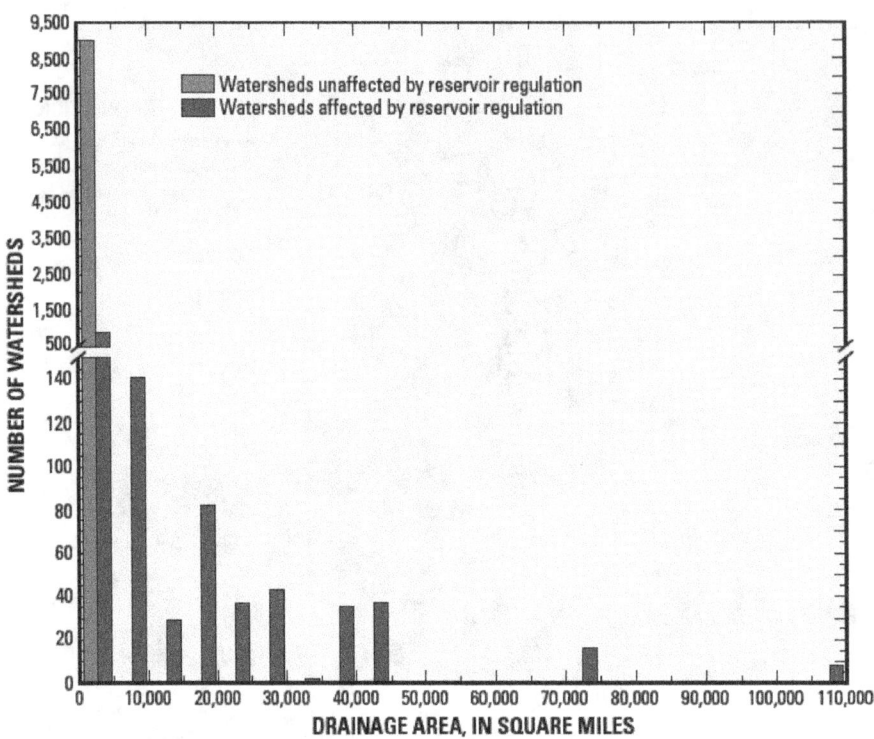

Figure 9.   Number of watersheds in the Upper Colorado River Basin by drainage area.

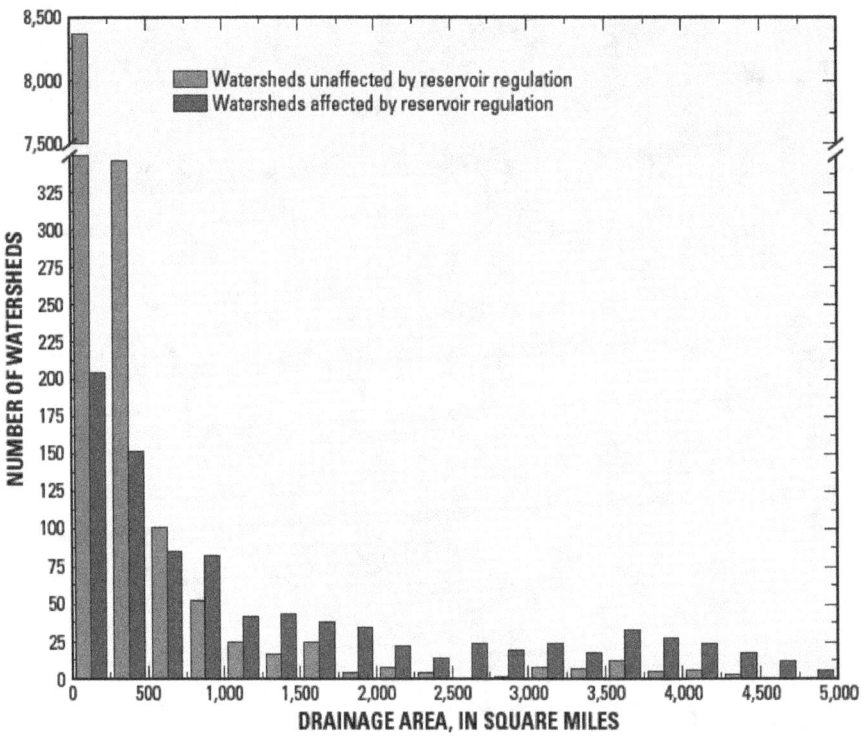

Figure 10.   Number of watersheds in the Upper Colorado River Basin with drainage areas less than 5,000 square miles.

gages to unregulated UCRB watersheds is at least 10 percent across the range of values of the watershed characteristic, then the USGS streamgage network is considered to represent that watershed characteristic. For the unregulated active USGS streamgage network, a criterion of 2 percent of the unregulated UCRB watersheds across the range of values for the watershed characteristic is used.

For regulated watersheds, the USGS streamgage network measures streamflow in about 17 percent of regulated UCRB watersheds (226 gages and 1,353 watersheds), and the active gaging station network measures streamflow in about 7.2 percent (98 gages and 1,353 watersheds) of the watersheds. Therefore, if the ratio of regulated USGS gages to regulated UCRB watersheds is at least 20 percent across the range of values for a specific watershed characteristic then the USGS streamgage network is considered to represent that watershed characteristic. The criterion of 10 percent is used for the ratio of regulated active USGS streamgages to regulated UCRB watersheds.

Watershed Characteristics

As discussed above, drainage area is a fundamental physical characteristic of a watershed that is commonly used to classify watersheds and is the basis from which many other watershed characteristics are calculated. For example, the percentage of a watershed covered by forest is computed in relation to the drainage area of the watershed; area covered by forest divided by the total area of the watershed. The discussion of watershed characteristics that follows in this section of the report is presented in terms of drainage area.

For this analysis, 16 watershed characteristics and one climatic characteristic were computed for the 1,053 watersheds with USGS streamgages and 10,338 watersheds in the UCRB using geographic information system (GIS) techniques (table 2). These characteristics include physiographic watershed attributes, such as mean basin elevation; land cover parameters, such as percentage of basin covered by evergreen forest; parameters that can assist in evaluating the anthropogenic influence upon basins, such as population density and

Table 2.   Watershed and climatic characteristics.

[NED, National Elevation Dataset; PRISM, Parameter-elevation Regressions on Independent Slopes Model; NLCD, National Land Cover Dataset]

Characteristic	Units	Datasets used
Mean basin elevation	Feet	NED
Mean basin average annual precipitation	Inches	PRISM 1971–2000 annual averages
Land Cover		
Area covered by developed land	Percent	2001 NLCD
Area covered by barren land	Percent	2001 NLCD
Area covered by deciduous forest	Percent	2001 NLCD
Area covered by evergreen forest	Percent	2001 NLCD
Area covered by mixed forest	Percent	2001 NLCD
Area covered by shrubs, young or stunted trees	Percent	2001 NLCD
Area covered by grass or herbaceous land	Percent	2001 NLCD
Lithologic classification		
Igneous and metamorphic	Percent	Geologic map of United States, 1:500,000-scale digital geologic maps of Arizona, Colorado, New Mexico, Utah, and Wyoming.
Sedimentary, basin fill (continental)	Percent	Geologic map of United States, 1:500,000-scale digital geologic maps of Arizona, Colorado, New Mexico, Utah, and Wyoming.
Sedimentary, carbonate (marine)	Percent	Geologic map of United States, 1:500,000-scale digital geologic maps of Arizona, Colorado, New Mexico, Utah, and Wyoming.
Sedimentary, clastic, Mesozoic	Percent	Geologic map of United States, 1:500,000-scale digital geologic maps of Arizona, Colorado, New Mexico, Utah, and Wyoming.
Sedimentary, clastic (continental), Tertiary	Percent	Geologic map of United States, 1:500,000-scale digital geologic maps of Arizona, Colorado, New Mexico, Utah, and Wyoming.
Sedimentary, mixed (continental and marine)	Percent	Geologic map of United States, 1:500,000-scale digital geologic maps of Arizona, Colorado, New Mexico, Utah, and Wyoming.
2000 population density	People per square mile	U.S. Geological Survey, 2000 population density by block group for the conterminous United States
Road density	Miles per square mile	U.S. Geological Survey, The National Map: Transportation

road density; and bedrock lithology. The watershed characteristics were selected to generally examine the physical, ecological, and climatological aspects of watersheds in the UCRB to assess how well the USGS streamgage network has monitored the spectrum of watersheds within the basin. The selected watershed characteristics are also widely used in hydrologic, biologic, and ecological studies of watersheds. The watersheds of the UCRB are further partitioned as either regulated or unregulated.

Elevation

Elevation in the UCRB ranges from 14,400 ft near the Continental Divide in Colorado to about 3,200 ft at Lees Ferry, Arizona. The mean elevation for the UCRB is 7,070 ft. Mean watershed elevation was computed for unregulated (fig. 11) and regulated (fig. 12) watersheds with USGS streamgages and watersheds in the UCRB.

The comparison between unregulated watersheds with USGS gages and UCRB watersheds for mean drainage basin elevation is shown in figure 11. In the scatter plot, unregulated watersheds with USGS gages are fairly evenly distributed across the range of drainage area and mean elevations of UCRB watersheds with one exception. A small gap in the entire USGS streamgage network exists for watersheds between 20 and 70 mi^2 with mean drainage basin elevations between 5,000 and 6,000 ft, indicating that the USGS streamgage network underrepresents watersheds in this mean elevation range and drainage area. Watersheds instrumented with active USGS gages are more common in watersheds with mean basin elevations greater than about 7,000 ft, indicating a bias in the active USGS streamgage network toward high elevation watersheds. Further, no less than 20 percent of the UCRB watersheds with mean basin elevations greater than 9,000 ft have been monitored by a USGS gaging station, whereas less than 5 percent of watersheds with a mean

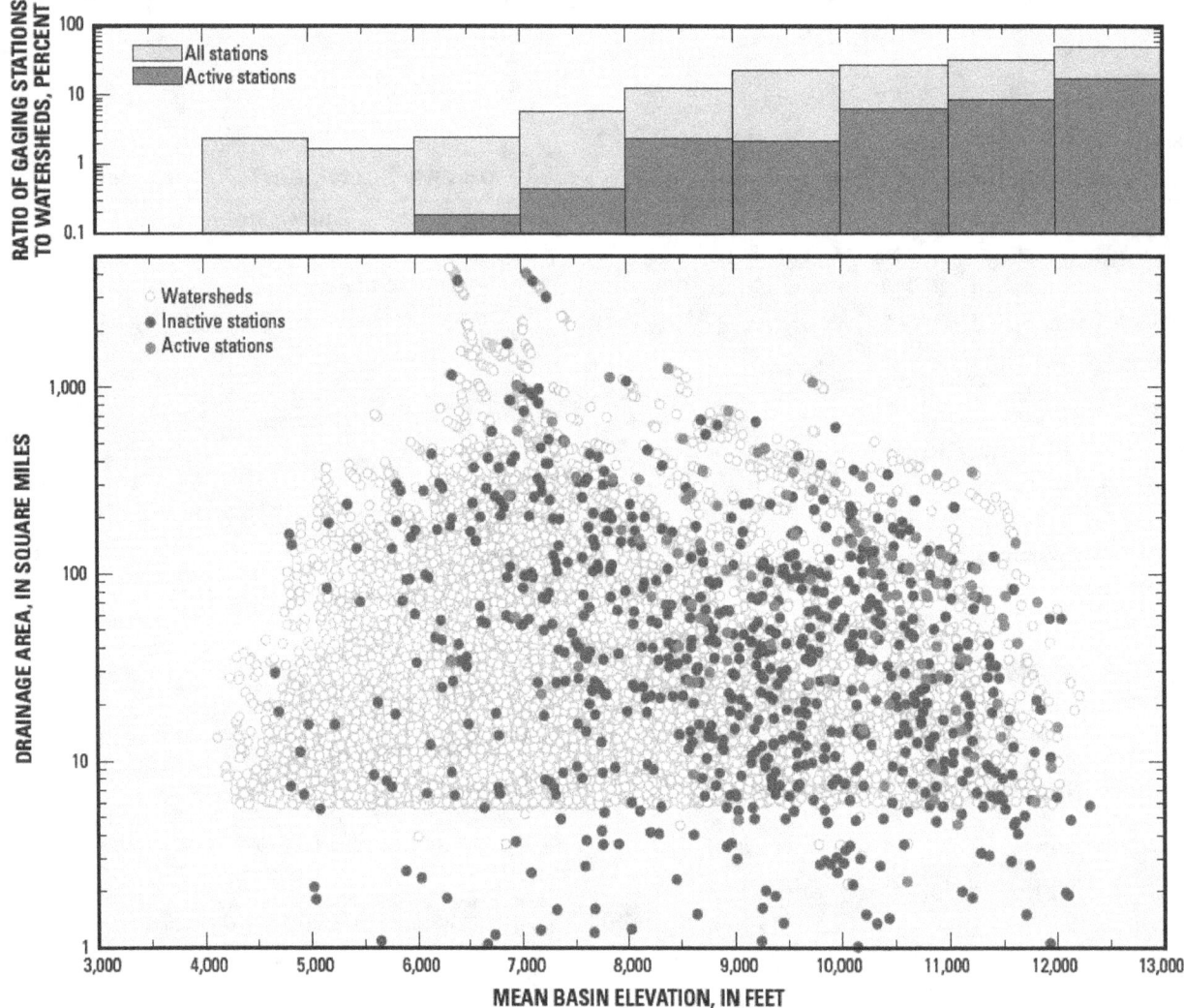

Figure 11.   Mean drainage basin elevation for watersheds and U.S. Geological Survey streamgage locations in the Upper Colorado River Basin that are unaffected by reservoir regulation.

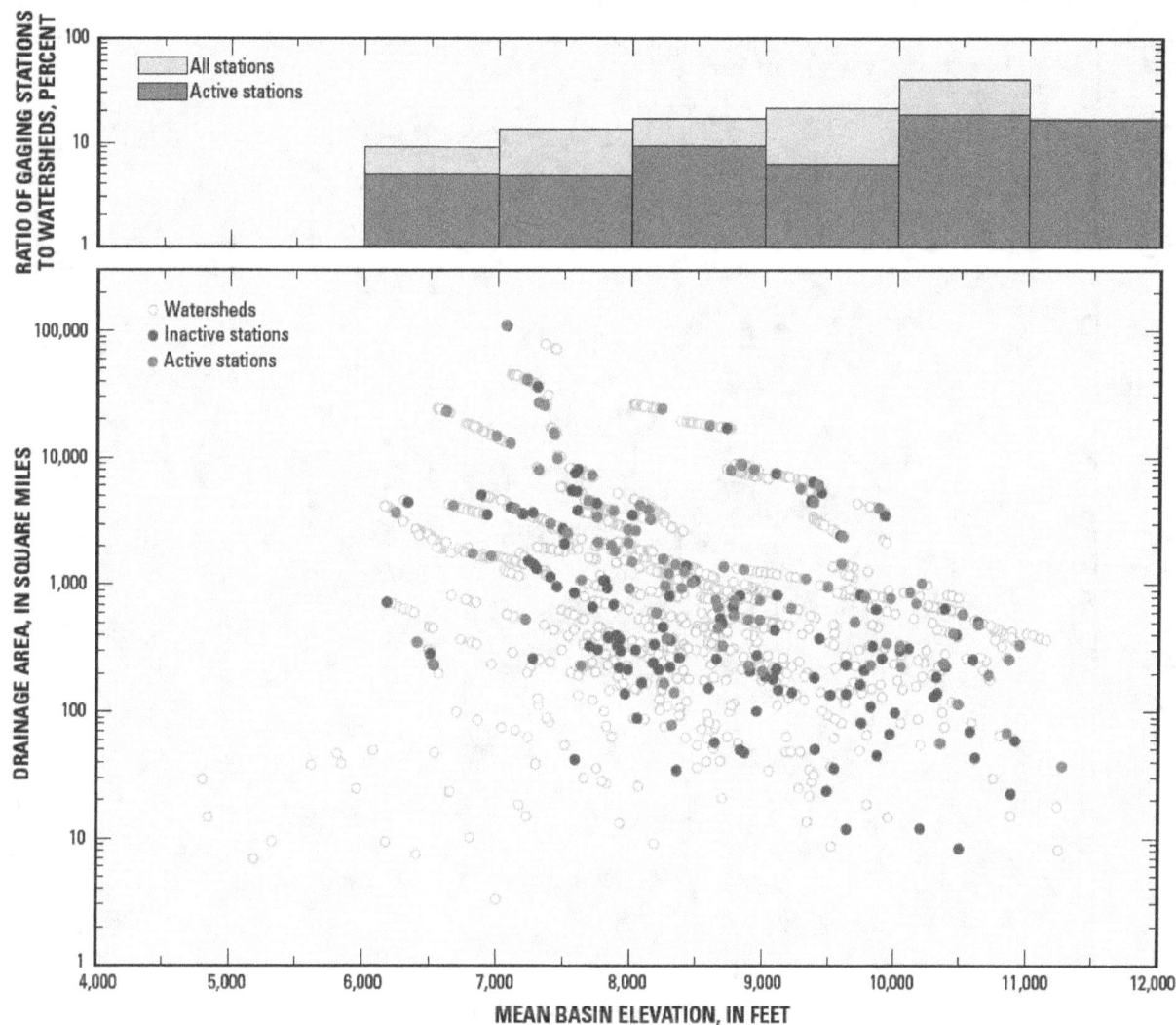

Figure 12. Mean drainage basin elevation for watersheds and U.S. Geological Survey streamgage locations in the Upper Colorado River Basin that are affected by reservoir regulation.

elevation of less than 8,000 ft have been monitored by a USGS gaging station. Therefore, small unregulated drainage basins with low mean basin elevations are not being represented well by the streamgage network in the UCRB.

Regulated watersheds typically have watershed areas greater than 100 mi[2] and mean basin elevations greater than 8,000 ft (fig. 12). The distribution of mean basin elevations for the regulated active USGS gages is similar to the distribution for the regulated inactive streamgage locations. Approximately 10 percent or more of the regulated UCRB watersheds with mean basin elevations greater than 6,000 ft have been monitored by a USGS streamgage and approximately 5 percent or more are being monitored currently.

Mean Annual Precipitation

The estimated annual precipitation in the UCRB ranges from about 50 in., mostly as snow, near the Continental Divide to about 10 in. on the Colorado Plateau. Mean watershed average annual precipitation was computed from PRISM (Precipitation-elevation Regressions on Independent Slopes Model) mean annual precipitation data for 1971–2000 (PRISM Group, Oregon State University, 2007) for unregulated (fig. 13) and regulated (fig. 14) UCRB watersheds and USGS streamgages.

On the basis of a criterion of 10 percent of unregulated UCRB watersheds with USGS streamgages and 2 percent of unregulated watersheds with active USGS streamgages, the USGS streamgage network and active USGS streamgages underrepresent UCRB watersheds having mean annual precipitation less than 25 in. For watersheds with an average annual precipitation greater than 25 in., the USGS streamgage

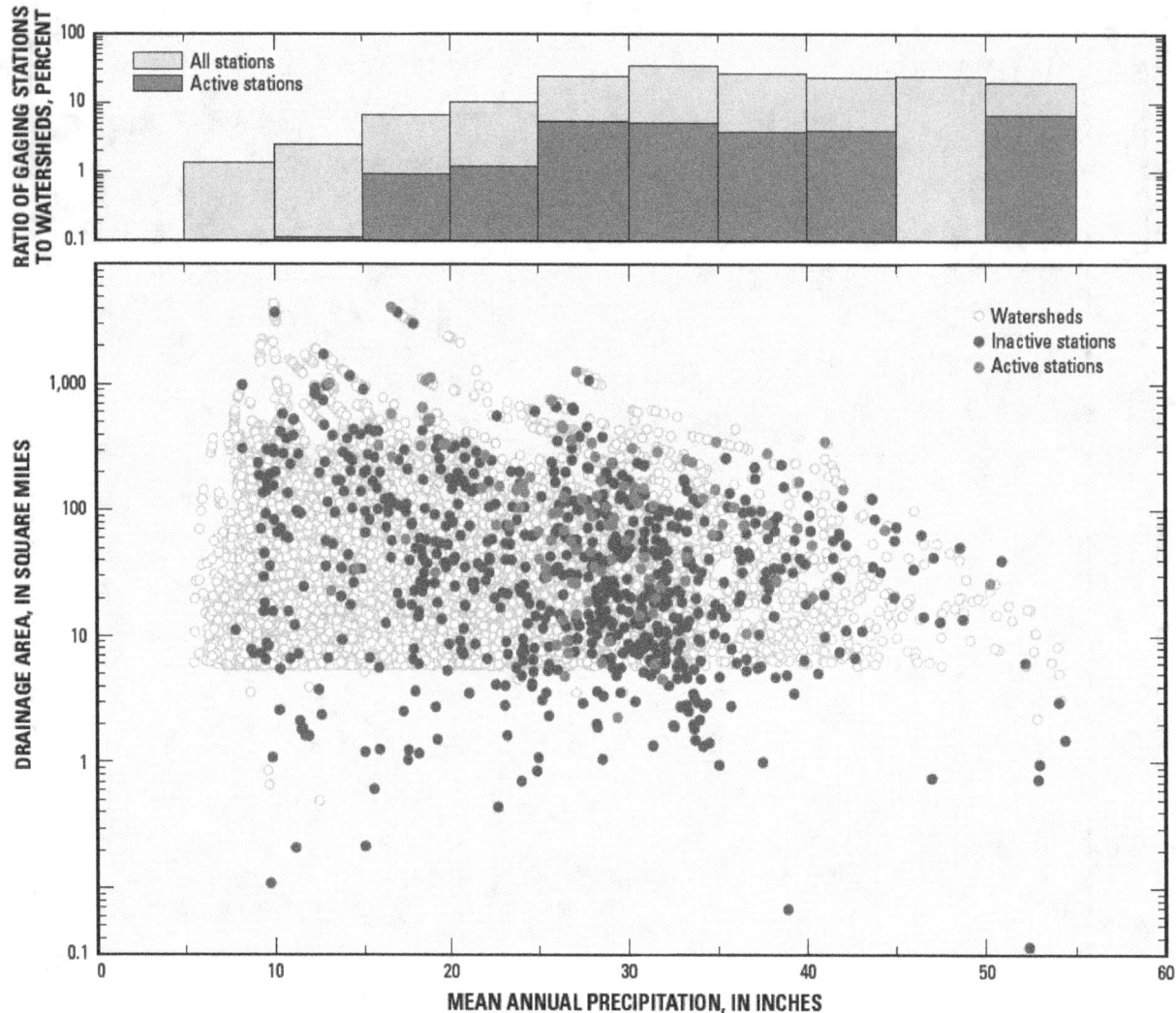

Figure 13.   Mean annual precipitation for watersheds and U.S. Geological Survey streamgages in the Upper Colorado River Basin that are unaffected by reservoir regulation.

network and active USGS gages exceed the criteria for representing the UCRB watersheds. This overrepresentation can be explained in part by the fact that in the UCRB, many watersheds at lower elevations with low mean annual precipitation have ephemeral streams. Ephemeral streams are less likely to be measured as part of the USGS streamgage network because there is less interest in low amounts of streamflow. Conversely, surface-water resources in the UCRB generally are associated with higher elevation areas having greater average annual precipitation. As a result, USGS streamgages are more likely to be located in these watersheds.

For regulated watersheds, both the entire USGS streamgage network and active USGS gages underrepresent UCRB watersheds having a mean annual precipitation of less than 20 in. (fig. 14). In contrast, watersheds with an average annual precipitation greater than 20 in. are represented well by the USGS streamgage network, including the active gages.

Land Cover

The vegetation and ecology of the UCRB vary primarily because of differences in elevation, latitude, and climate. Because it is important to take ecological issues into account as the region develops, the extent to which the USGS streamgage network represents these various landscapes is important for managing water, land, and ecological resources. The 2001 National Land Cover Dataset (NLCD; Homer and others, 2004) provides a geospatial classification of land cover for the Nation. The NLCD classes of land cover in the UCRB used in this assessment include developed land, barren land, deciduous forest, evergreen forest, mixed forest, shrub-scrub land, and grass or herbaceous land (fig. 15). These land cover classes are described in table 3. The developed lands class is a sum of the four developed land classifications in the NLCD: developed open space, developed low intensity, developed medium intensity, and developed high intensity.

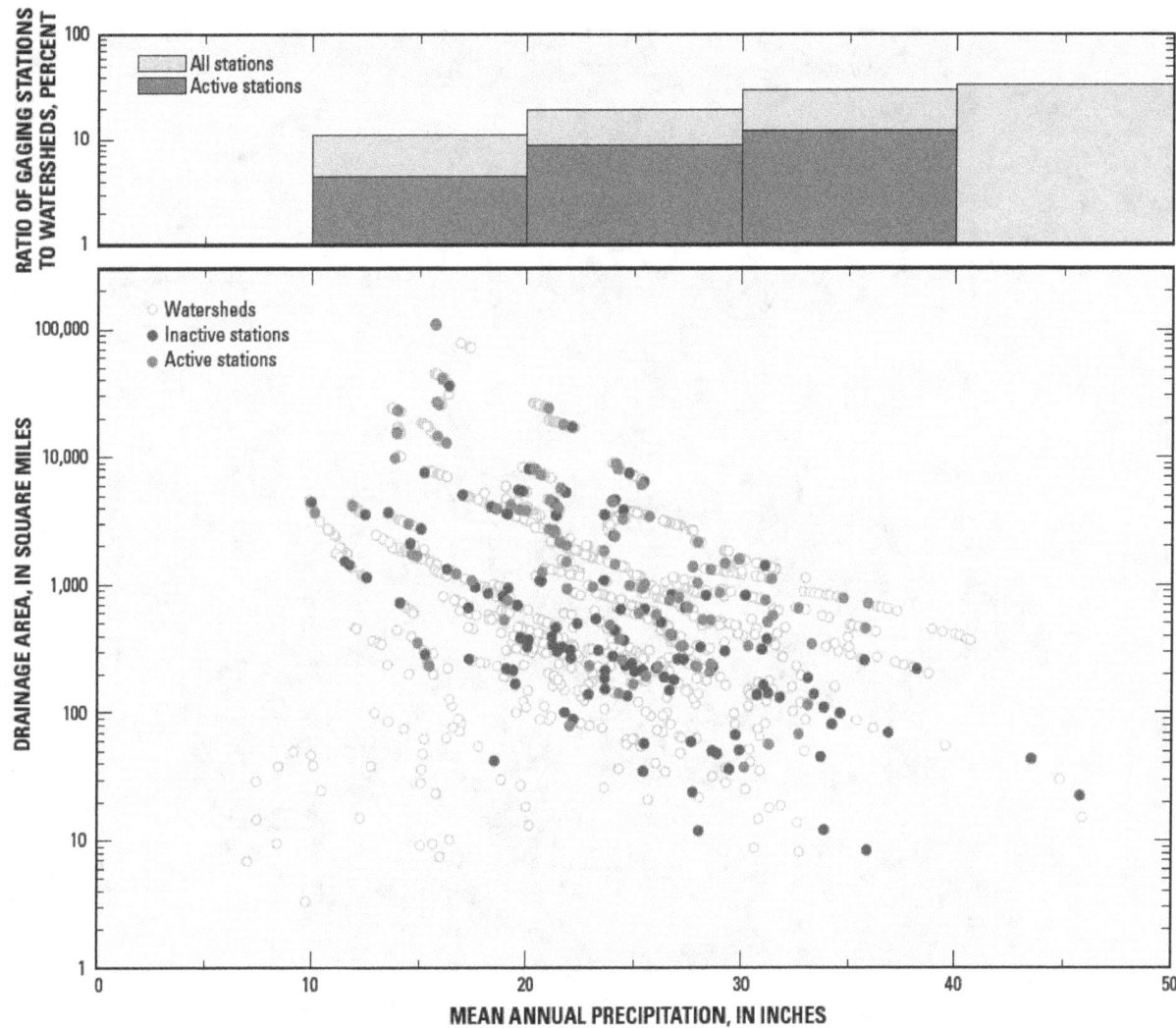

Figure 14. Mean annual precipitation for watersheds and U.S. Geological Survey streamgages in the Upper Colorado River Basin that are affected by reservoir regulation.

Table 3. Description of land cover classifications from the National Land Cover Dataset.

Developed: Areas characterized by a high percentage (30 percent or greater) of constructed materials (e.g., asphalt, concrete, buildings, etc.).

Barren Land (Rock/Sand/Clay): Barren areas of bedrock, desert pavement, scarps, talus, slides, volcanic material, glacial debris, sand dunes, strip mines, gravel pits and other accumulations of earthen material. Generally, vegetation accounts for less than 15 percent of total cover.

Deciduous Forest: Areas dominated by trees generally greater than 5 meters tall and composing greater than 20 percent of total vegetation cover. More than 75 percent of the tree species shed foliage simultaneously in response to seasonal change.

Evergreen Forest: Areas dominated by trees generally greater than 5 meters tall and composing greater than 20 percent of total vegetation cover. More than 75 percent of the tree species maintain their leaves all year. Canopy is never without green foliage.

Mixed Forest: Areas dominated by trees generally greater than 5 meters tall and composing greater than 20 percent of total vegetation cover. Neither deciduous nor evergreen species are greater than 75 percent of total tree cover.

Shrub/Scrub: Areas dominated by shrubs less than 5 meters tall with shrub canopy typically composing greater than 20 percent of total vegetation. This class includes true shrubs, young trees in an early successional stage, or trees stunted from environmental conditions.

Grassland/Herbaceous: Areas dominated by grammanoid or herbaceous vegetation generally composing greater than 80 percent of total vegetation. These areas are not subject to intensive management such as tilling, but can be utilized for grazing.

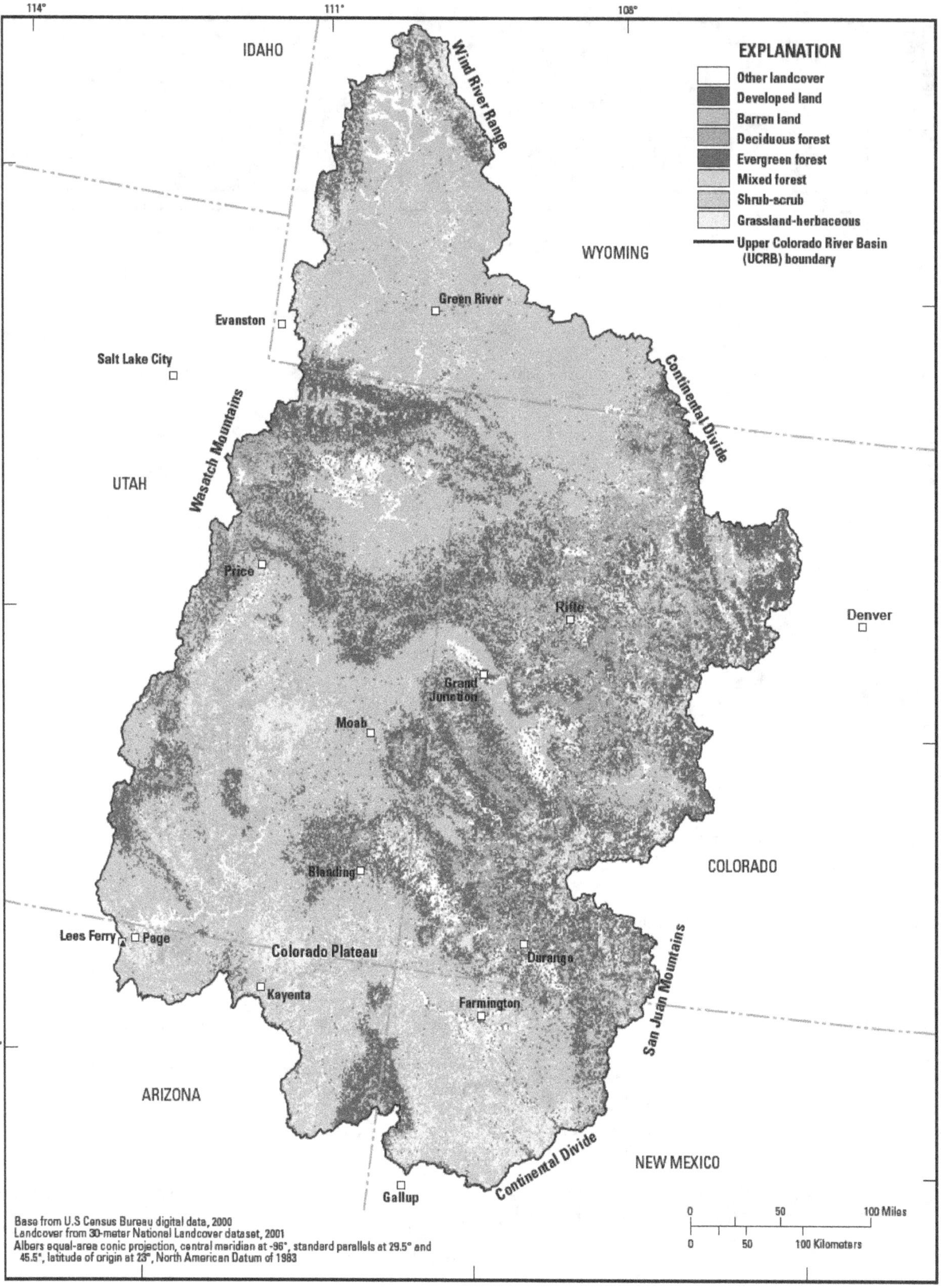

EXPLANATION

Other landcover
Developed land
Barren land
Deciduous forest
Evergreen forest
Mixed forest
Shrub-scrub
Grassland-herbaceous
Upper Colorado River Basin
(UCRB) boundary

IDAHO

Wind River Range

WYOMING

Continental Divide

Evanston

Green River

Salt Lake City

Wasatch Mountains

UTAH

Price

Rifle

Denver

Grand Junction

Moab

Blanding

COLORADO

Lees Ferry   Page

Colorado Plateau

Durango

San Juan Mountains

Kayenta

Farmington

ARIZONA

Continental Divide

NEW MEXICO

Gallup

Base from U.S Census Bureau digital data, 2000
Landcover from 30-meter National Landcover dataset, 2001
Albers equal-area conic projection, central meridian at -96°, standard parallels at 29.5° and
45.5°, latitude of origin at 23°, North American Datum of 1983

0   50   100 Miles

0   50   100 Kilometers

Figure 15.   Distribution of seven land cover classifications from the National Land Cover Dataset for the Upper Colorado River Basin.

The percentage of the drainage area covered by each of the seven land classifications was computed for unregulated and regulated UCRB watersheds both instrumented and not instrumented with USGS streamgages (figs. 16 through 29). The most prevalent land cover in the UCRB is shrub-scrub, which accounts for about 50 percent of the land cover. Conversely, lands classified as developed compose about 1 percent of the total land cover in the UCRB. The land cover characteristics, as well as their distribution in watersheds of various sizes in the UCRB, are presented in figures 16 through 29.

For unregulated watersheds, the USGS streamgage network generally is not evenly distributed across the drainage basin area for the land cover classes, except in the cases of deciduous and evergreen forest land cover (figs. 16 and 17). USGS streamgages, including active gages, are not as evenly distributed across drainage basin size and percent of land cover for the other five land cover classes in unregulated watersheds as the forest cover types (figs. 18–22). For the

shrub-scrub class, the most prevalent land cover class in the UCRB, the USGS streamgage network usually better represents watersheds with less than 20 percent shrub-scrub land cover than ones with greater shrub-scrub coverage (fig. 18). This could be because watersheds dominated by shrub-scrub are at lower elevations and tend to be drier, which would suggest that there are fewer perennial streams. This same pattern and explanation can apply to the barren land and grassland classes as well (figs. 19 and 20). The biases visible in the graphs most likely are a function of the distribution of water within the UCRB and thus, within the watershed dataset. It is important to understand that although a watershed has been delineated, this does not mean there is perennial flow in the watershed to measure. For unregulated watersheds, the USGS streamgage network is more representative of the UCRB landscape when the drier land cover classes (shrub-scrub, grassland, and barren land) compose relatively small percentages of the total land cover.

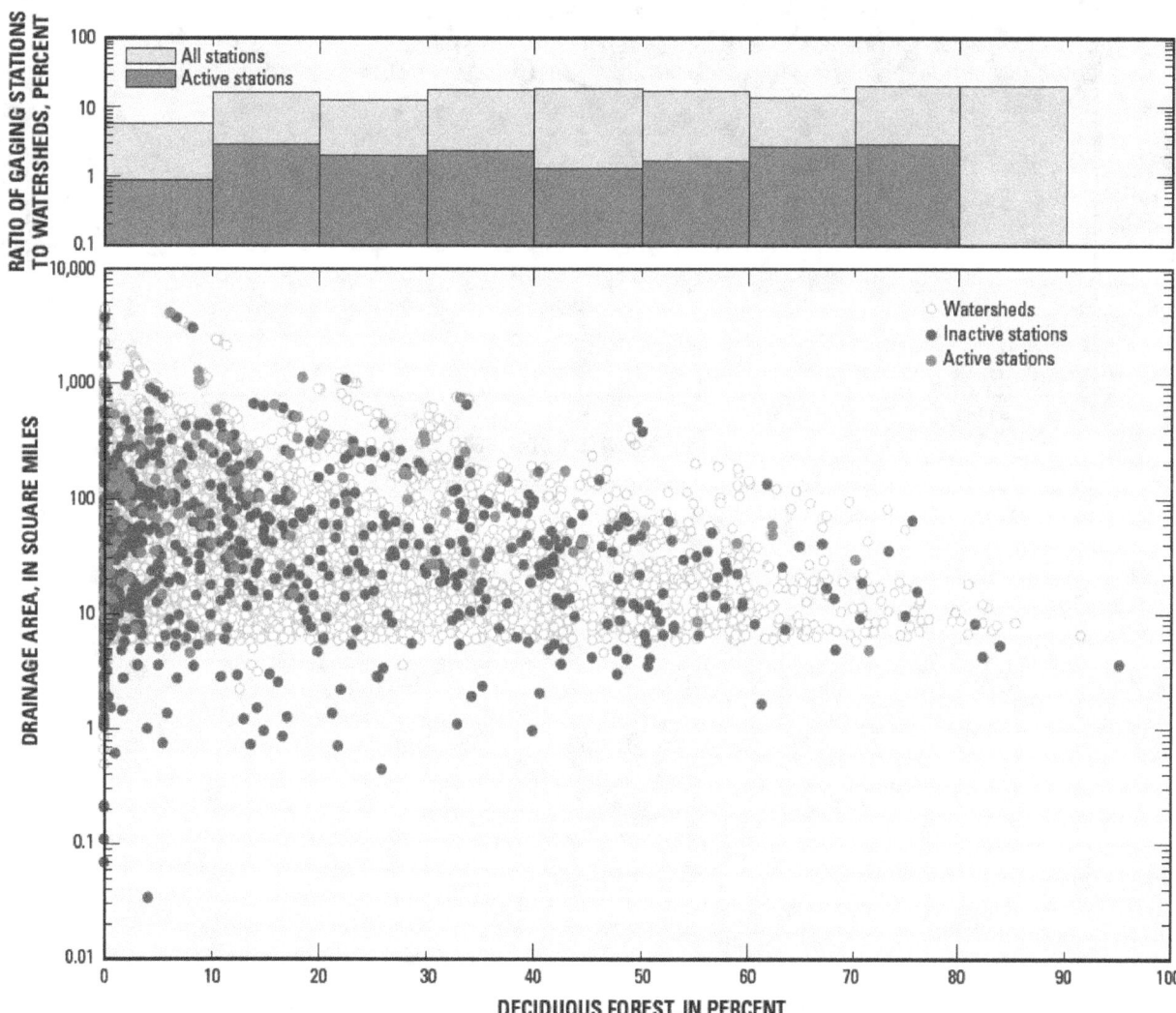

Figure 16. Percentage of drainage area covered by deciduous forest for watersheds and U.S. Geological Survey streamgages in the Upper Colorado River Basin that are unaffected by reservoir regulation.

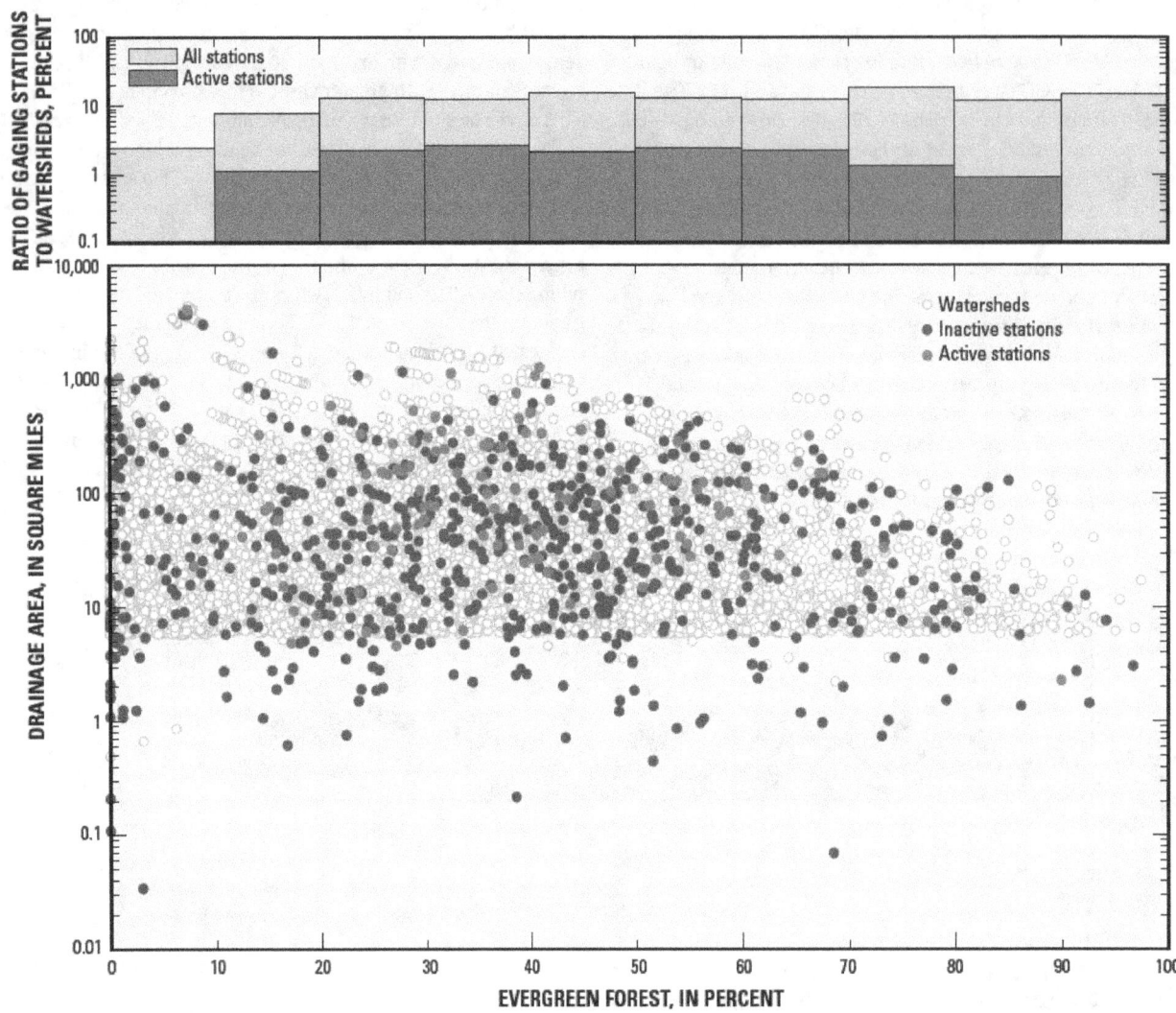

Figure 17. Percentage of drainage area covered by evergreen forest for watersheds and U.S. Geological Survey streamgages in the Upper Colorado River Basin that are unaffected by reservoir regulation.

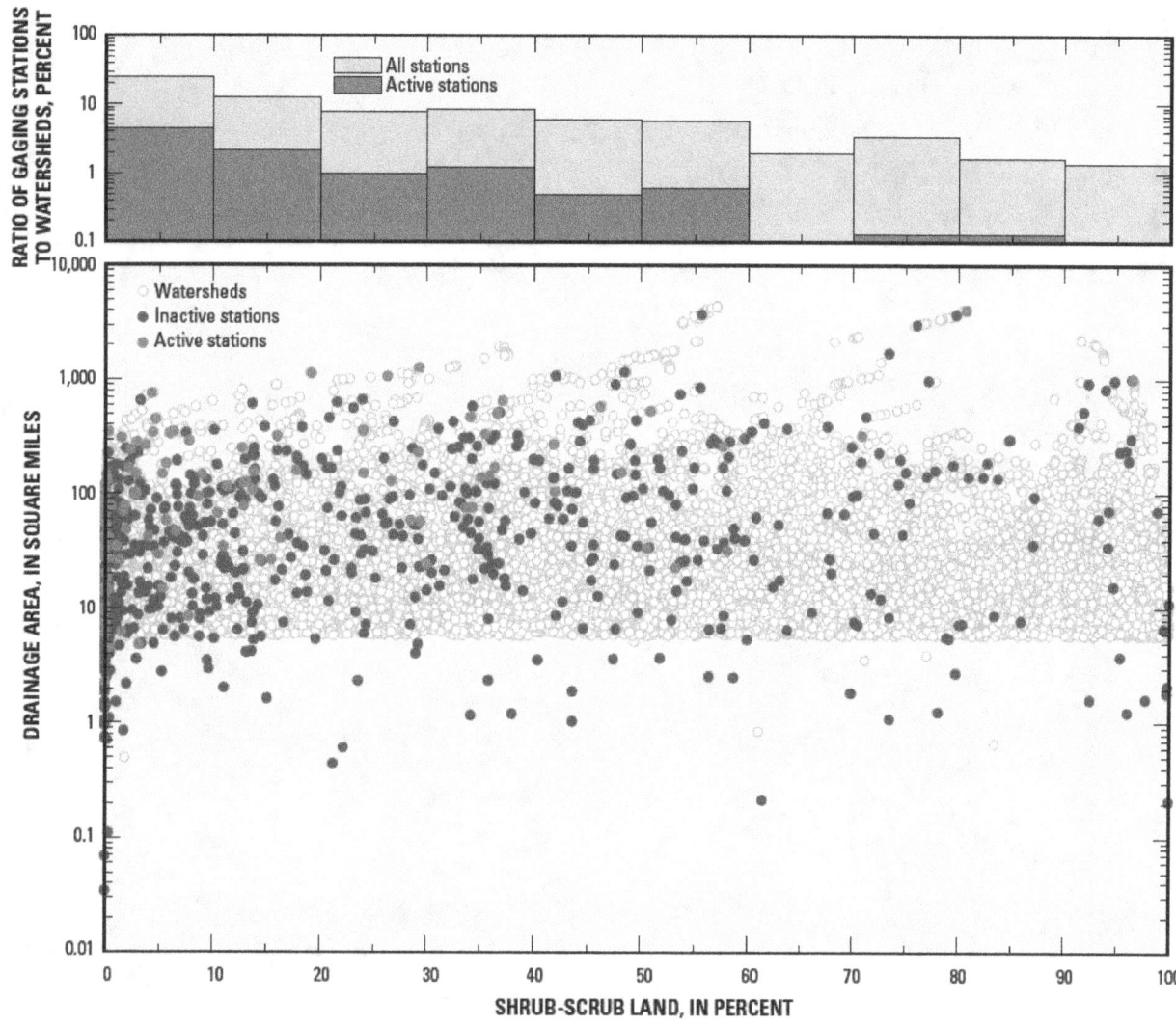

Figure 18. Percentage of drainage area covered by shrubs, young or stunted trees, for watersheds and U.S. Geological Survey streamgages in the Upper Colorado River Basin that are unaffected by reservoir regulation.

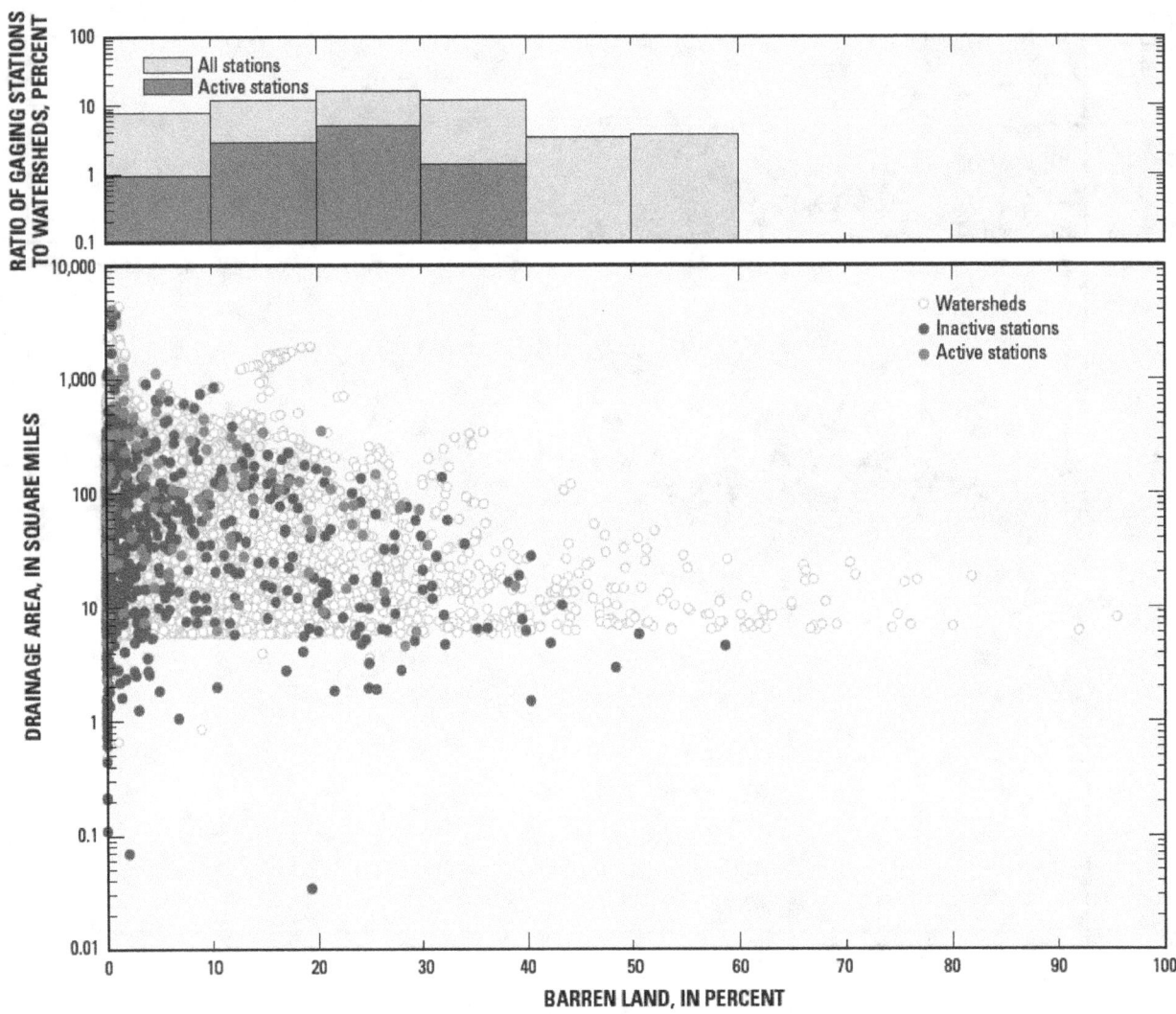

Figure 19.   Percentage of drainage area covered by barren land for watersheds and U.S. Geological Survey streamgages in the Upper Colorado River Basin that are unaffected by reservoir regulation.

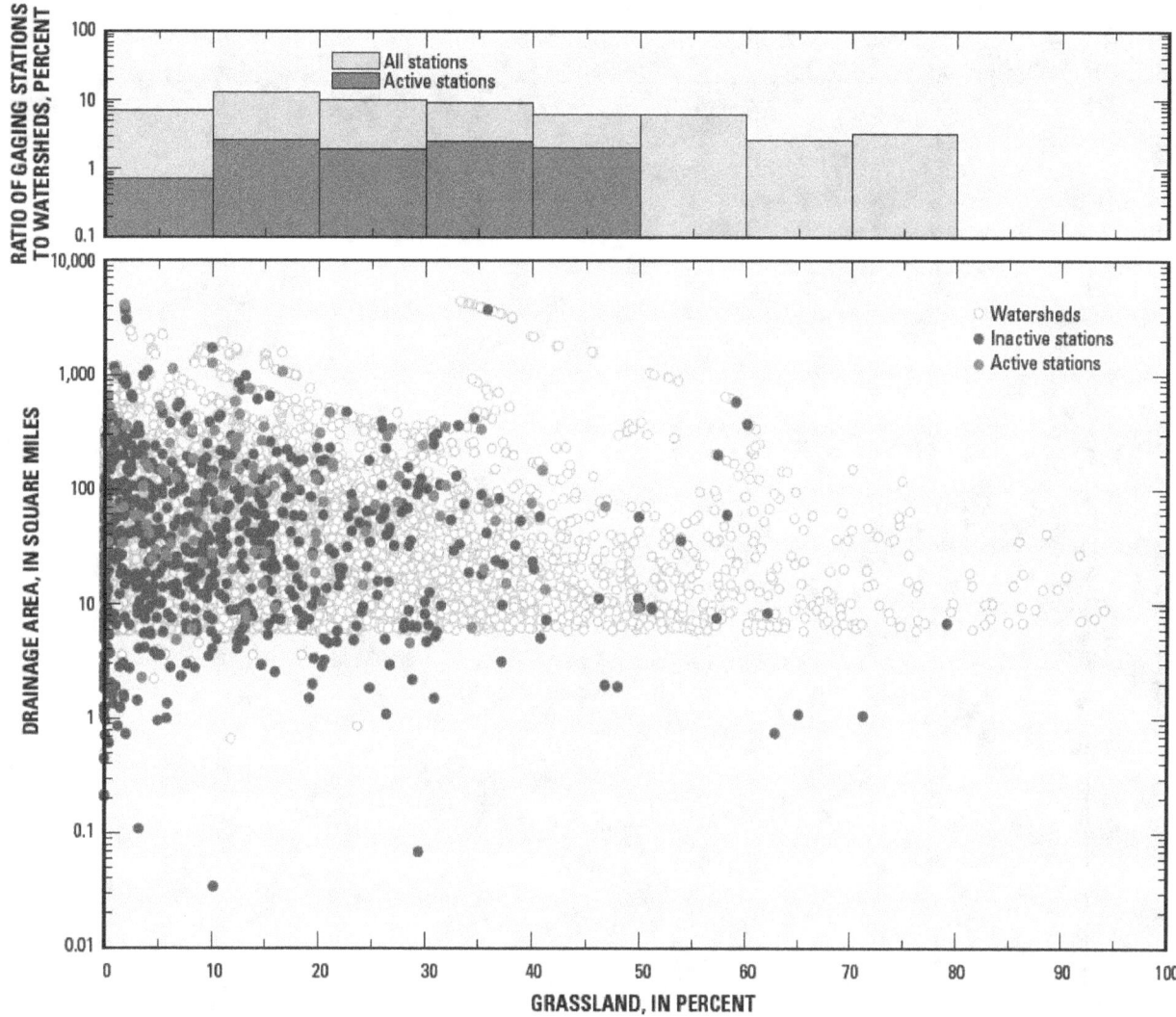

Figure 20. Percentage of drainage area covered by grass or herbaceous land for watersheds and U.S. Geological Survey streamgages in the Upper Colorado River Basin that are unaffected by reservoir regulation.

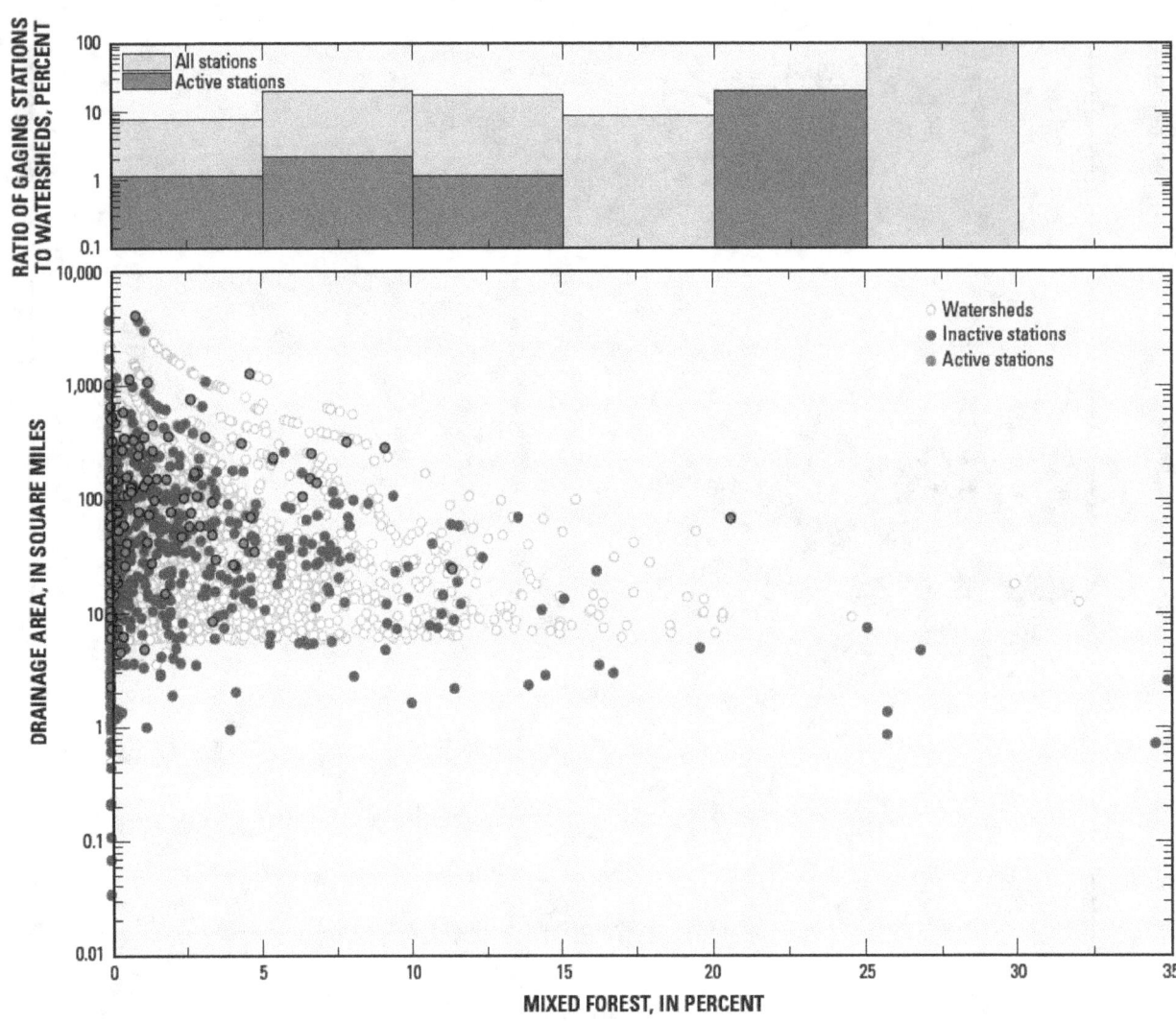

Figure 21.   Percentage of drainage area covered by mixed forest for watersheds and U.S. Geological Survey streamgages in the Upper Colorado River Basin that are unaffected by reservoir regulation.

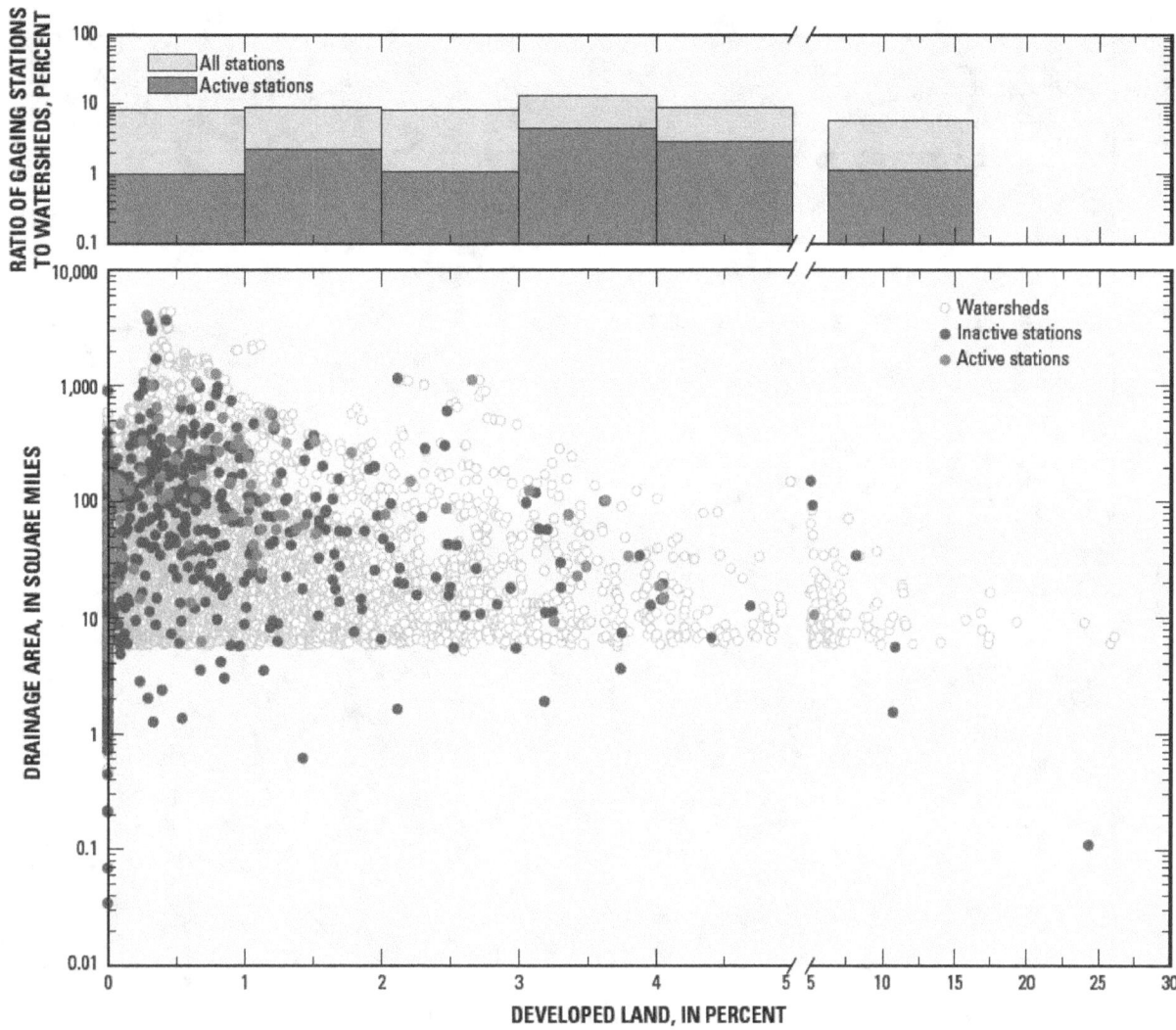

Figure 22. Percentage of drainage area covered by developed land for watersheds and U.S. Geological Survey streamgages in the Upper Colorado River Basin that are unaffected by reservoir regulation.

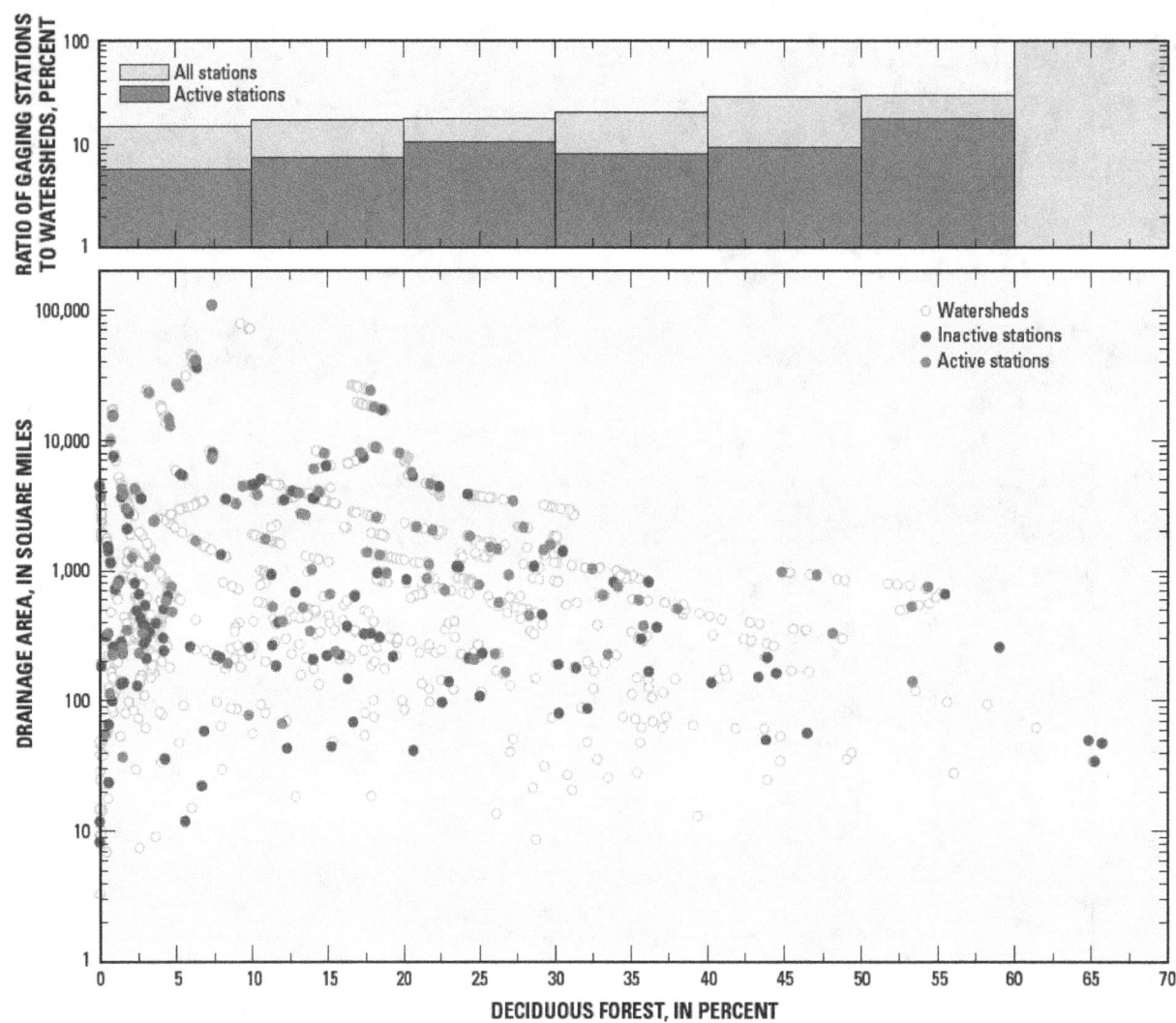

Figure 23.   Percentage of drainage area covered by deciduous forest for watersheds and U.S. Geological Survey streamgages in the Upper Colorado River Basin that are affected by reservoir regulation.

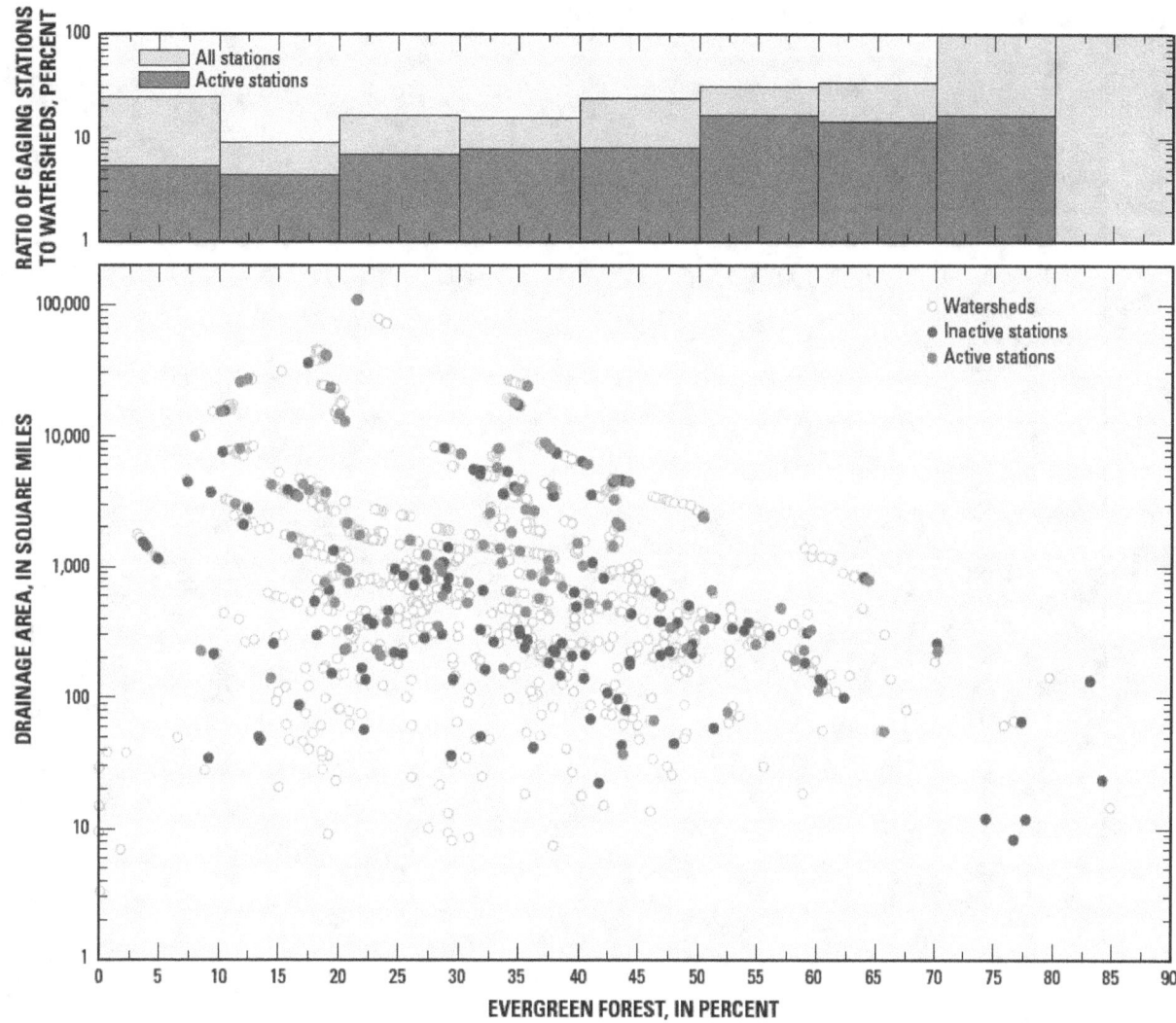

Figure 24. Percentage of drainage area covered by evergreen forest for watersheds and U.S. Geological Survey streamgages in the Upper Colorado River Basin that are affected by reservoir regulation.

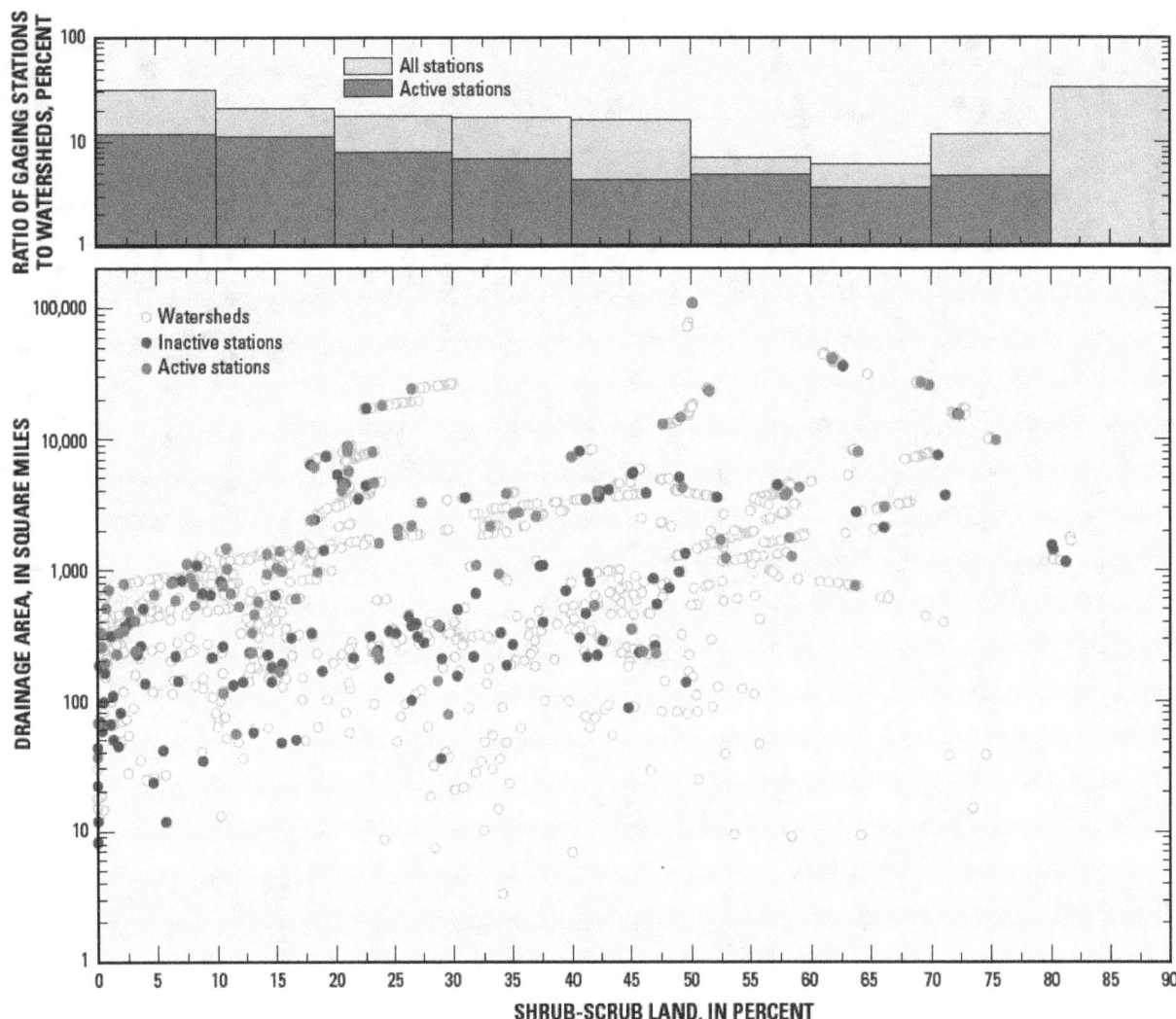

Figure 25.   Percentage of drainage area covered by shrubs, young or stunted trees for watersheds and U.S. Geological Survey streamgages in the Upper Colorado River Basin that are affected by reservoir regulation.

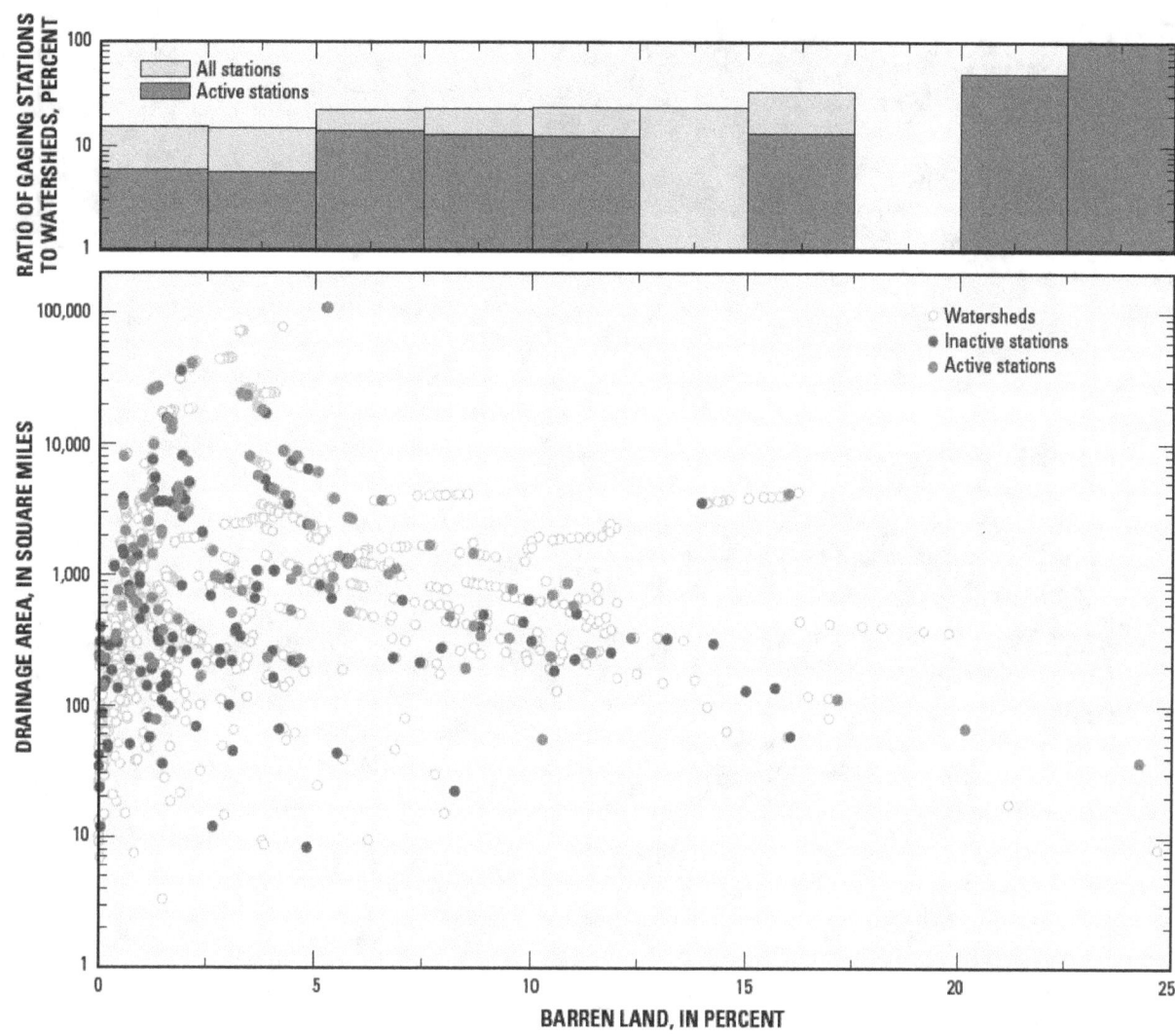

Figure 26. Percentage of drainage area covered by barren land for watersheds and U.S. Geological Survey streamgages in the Upper Colorado River Basin that are affected by reservoir regulation.

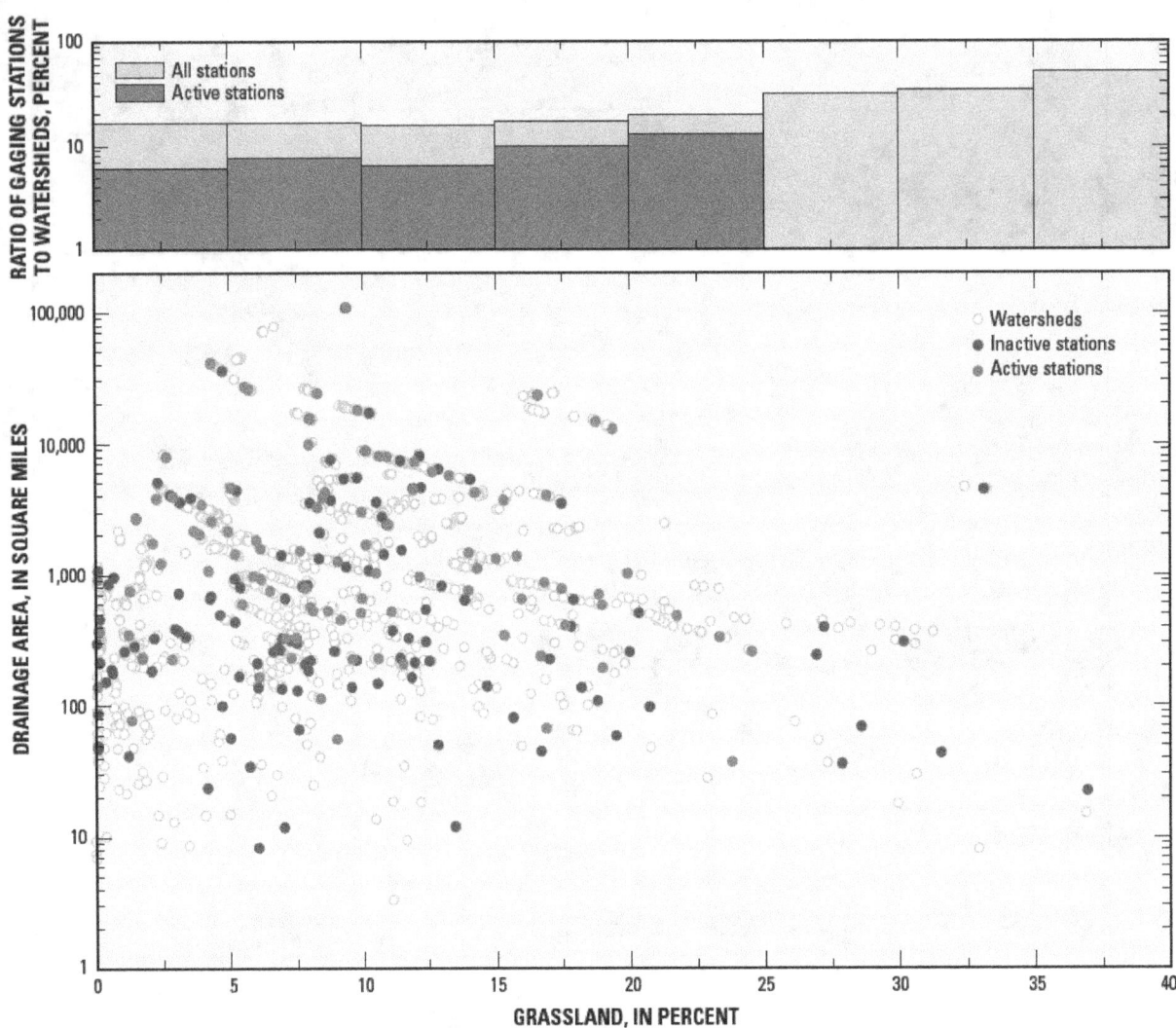

Figure 27.   Percentage of drainage area covered by grass or herbaceous land for watersheds and U.S. Geological Survey streamgages in the Upper Colorado River Basin that are affected by reservoir regulation.

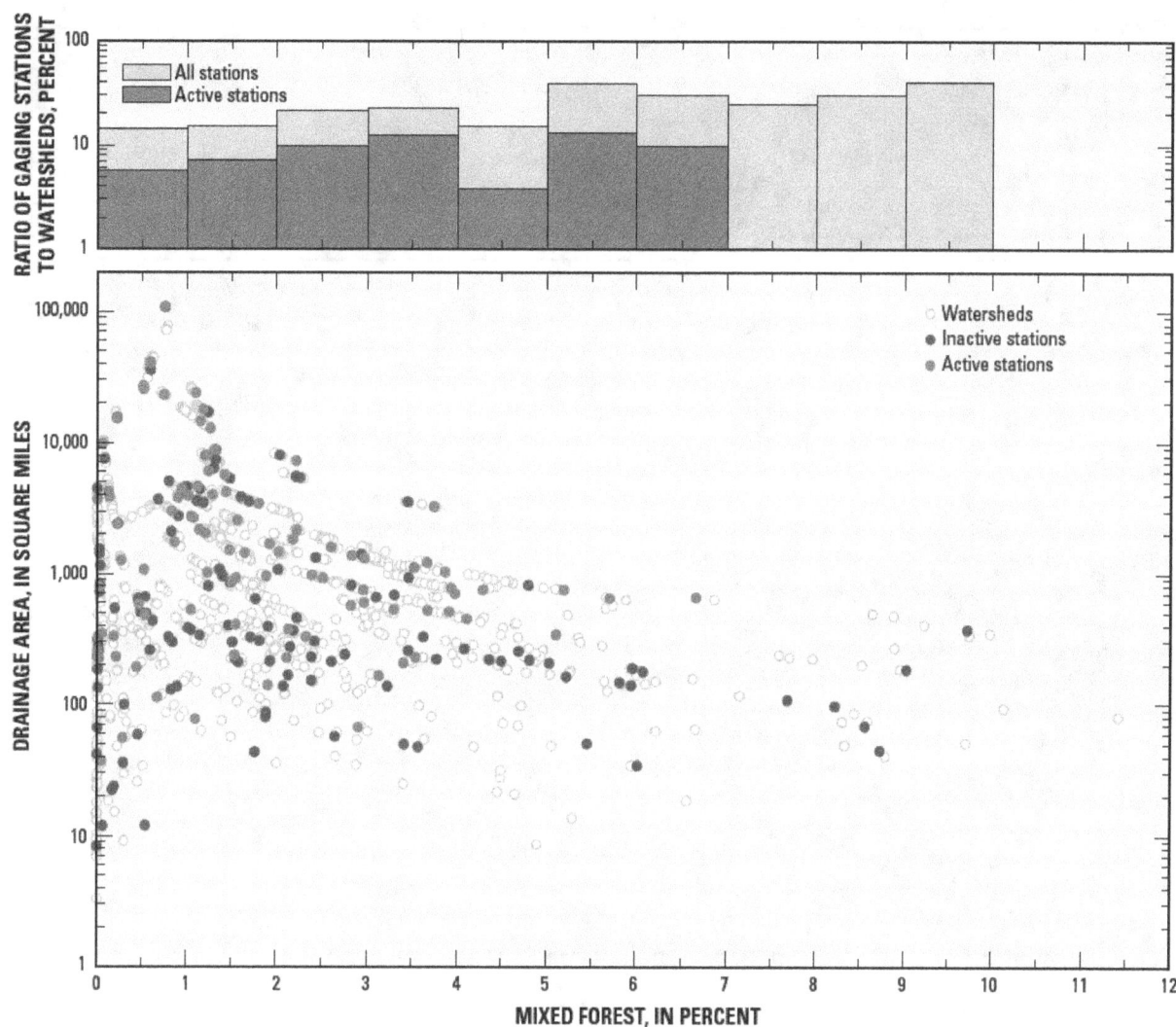

Figure 28. Percentage of drainage area covered by mixed forest for watersheds and U.S. Geological Survey streamgages in the Upper Colorado River Basin that are affected by reservoir regulation.

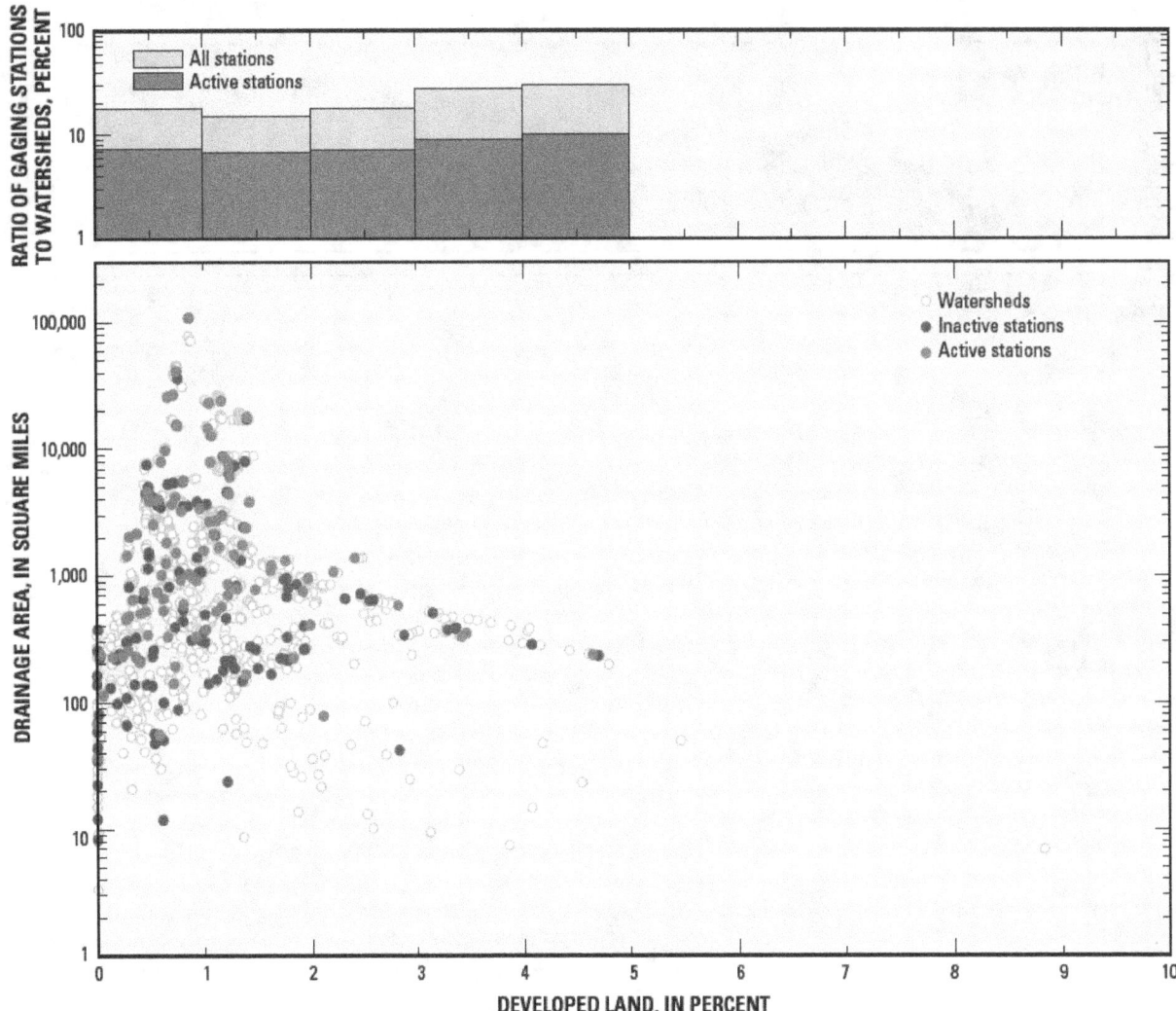

Figure 29.   Percentage of drainage area covered by developed land for watersheds and U.S. Geological Survey streamgages in the Upper Colorado River Basin that are affected by reservoir regulation.

Similarly, regulated watersheds instrumented with USGS streamgages tend to represent well the deciduous and evergreen forest classes for UCRB watersheds (figs. 23 and 24), although watersheds with active USGS gages slightly underrepresent UCRB watersheds for these two land cover classes. Watersheds with USGS streamgages, including active USGS gages, for the remaining five land cover classes underrepresent UCRB watersheds, with the exception of the developed land class (fig. 29). For the shrub-scrub land cover class, watersheds with greater than 50 percent shrub-scrub land cover are underrepresented (fig. 25).

Bedrock Lithology

The geology of the UCRB is diverse, encompassing the Wind River Mountains in Wyoming, the San Juan Mountains in Colorado, and the Uinta Mountains in Utah. The undeformed rocks of the Colorado Plateau, which generally covers

the lower one-third of the UCRB, stand in distinct contrast to the highly deformed crustal blocks of the surrounding mountains. As part of this assessment of the USGS streamgage network, the bedrock geology of UCRB watersheds and watersheds with USGS streamgages was mapped and compared to evaluate how well the USGS streamgaging network represents the bedrock geology of watersheds in the UCRB. Six distinct lithologic classifications of bedrock (table 4) from a simplified 1:500,000 geologic map of the UCRB (Kenney and others, 2009) were developed (fig. 30). Because there are less than 9 mi^2 of sedimentary, basin-fill deposits (continental) in the UCRB, this lithology was not evaluated. The percentage of the drainage area for each of the remaining five lithologies was computed for unregulated and regulated watersheds with USGS streamgages and UCRB watersheds (figs. 31–40).

The lithologic classes sedimentary, clastic-Tertiary, and sedimentary, clastic-Mesozoic, account for about 80 percent of the bedrock geology in the UCRB, and the other lithologic classes account for about 5 to 10 percent each. In unregulated watersheds for these two lithologic classes, the USGS streamgage network and the active USGS network are representative of the watersheds in the UCRB, with the exception of watersheds having greater than 80 percent of either of these two classes (figs. 31 and 32). It is likely that these watersheds are within the Colorado Plateau at lower elevations, where there are few perennial streams and thus, fewer USGS gages. The USGS streamgage network and active USGS network underrepresent unregulated UCRB watersheds with small percentages (less than 10 percent) of the other three lithologic classes (figs. 33–35). In addition, the active USGS network does not represent the watersheds well at a variety of percentages for these three lithologic classes, indicating gaps in the active streamgage network. The USGS streamgage network and active USGS network exceed the criteria for representing the UCRB watersheds having 90–100 percent igneous and metamorphic lithologies. Again, this can be a function of where perennial streams are located as many of these watersheds are dominated by snowmelt runoff at high elevations and thus, are more likely to contain perennial streamflow.

For regulated watersheds, the USGS streamgage network and active USGS network adequately represent UCRB watersheds for all lithologic classes, with the exception of the sedimentary, mixed (continental and marine) lithology. The active USGS gages were not representative of watersheds (figs 36–40) having 30–40 percent and 50–80 percent of the sedimentary, mixed (continental and marine) lithology (fig. 39).

Population Density

Anthropogenic activities in a watershed can cause changes in streamflow patterns, timing, and volume. For example, land development affects the natural rainfall-runoff relation by changing the natural infiltration characteristics of the watershed. This often leads to increased surface runoff and streamflow, as well as increases in the rate at which runoff moves through the watershed. For this assessment, population density was used as a surrogate measure of anthropogenic activity upon watersheds in the UCRB. The 2000 population density (Hitt, 2003) was computed for unregulated and

regulated UCRB watersheds and watersheds instrumented with USGS streamgages (figs. 41 and 42).

The UCRB is largely a rural region that has a population density of 1.8 people/mi^2. Much of the land development is associated with agriculture rather than urban or suburban development. The population density of unregulated watersheds is skewed toward low population densities. The USGS streamgage network and the active USGS gages slightly underrepresent UCRB unregulated watersheds for population densities less then 15 people/mi^2. Unregulated UCRB watersheds where population density is greater than 15 people/mi^2 are not represented well in the USGS streamgage network. In addition, there are no regulated USGS streamgages in watersheds with populations greater than 10 people/mi^2. This shows that the USGS streamgage network underrepresents high population density watersheds relative to the UCRB. Overall, for regulated watersheds, the criteria of 20 percent coverage for the entire USGS network and 10 percent for the active USGS network are rarely met for population density.

Road Density

The density of roads in a watershed also can provide an indication of the amount of anthropogenic activity in a watershed. Similar to development, roads have a direct influence on the runoff characteristics of a watershed. Roads are often constructed of impervious materials such as concrete or asphalt, or can be simply bare ground. Surface runoff from all road types is generally greater than natural conditions because of the impervious nature of the material, the compact condition of the road bed, and (or) the lack of vegetation compared with the surrounding landscape. The density of all road types, in mi/mi^2, was computed from mapped roads (U.S. Geological Survey, 2010) for unregulated and regulated watersheds in the UCRB as well as those instrumented with USGS streamgages (figs. 43 and 44). The average road density for the entire UCRB is 1.6 mi/mi^2. Computed road densities can be considered a minimum road density because the mapped roads' dataset often does not include many smaller and minor, dirt and gravel roads. This is particularly true in areas of recent energy development and in areas where recreational roads have grown and propagated (Buto and others, 2010).

The UCRB is a rural region, thus, road densities are low, ranging from 0 to about 9 mi/mi^2. For unregulated watersheds, the USGS streamgage network and the active USGS gages are representative of UCRB watersheds having road densities less than 2 mi/mi^2, but the USGS streamgage network underrepresents UCRB watersheds with road densities greater than 2 mi/mi^2. Many unregulated watersheds in the UCRB have road densities greater than 3 mi/mi^2, yet there are only 19 gages, 3 of which are active, in watersheds where road densities exceed 3 mi/mi^2. Regulated watersheds where roads densities are less than 1.5 mi/mi^2 are represented well by the USGS streamgage network, including the active gages, but watersheds where road densities are greater than 1.5 mi/mi^2 are underrepresented by the USGS streamgage network.

Table 4. Six lithologic classifications in the Upper Colorado River Basin (Kenney and others, 2009).

Lithologic classifications	Total area (square miles)
Sedimentary, clastic, Mesozoic	50,200
Sedimentary, clastic (continental), Tertiary	38,000
Igneous and metamorphic	9,660
Sedimenatry, mixed (continental and marine)	5,910
Sedimentary, carbonate (marine)	4,240
Sedimentary, basin-fill (continental)	8.91

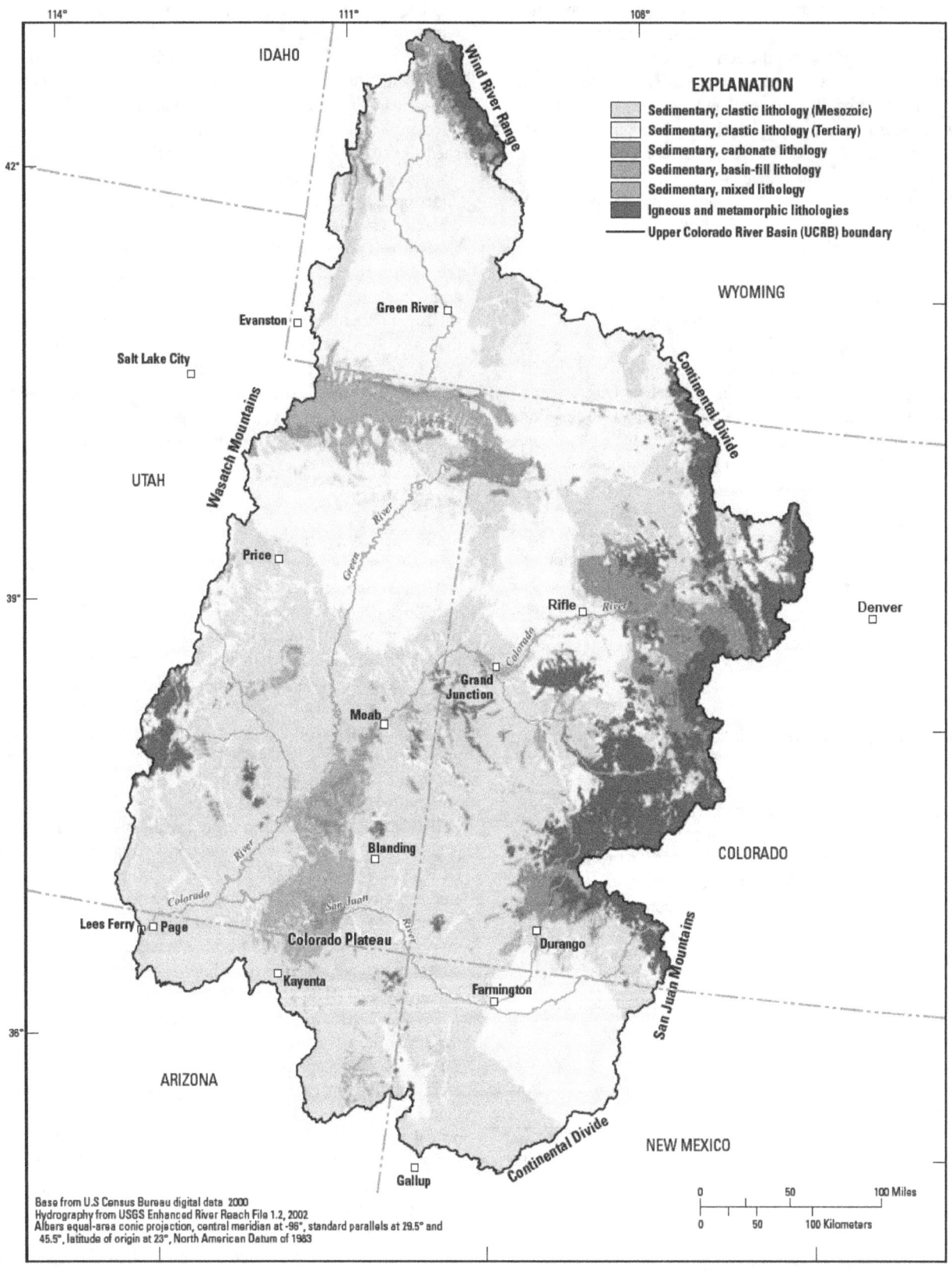

Figure 30.   Distribution of the six lithologic classifications in the Upper Colorado River Basin.

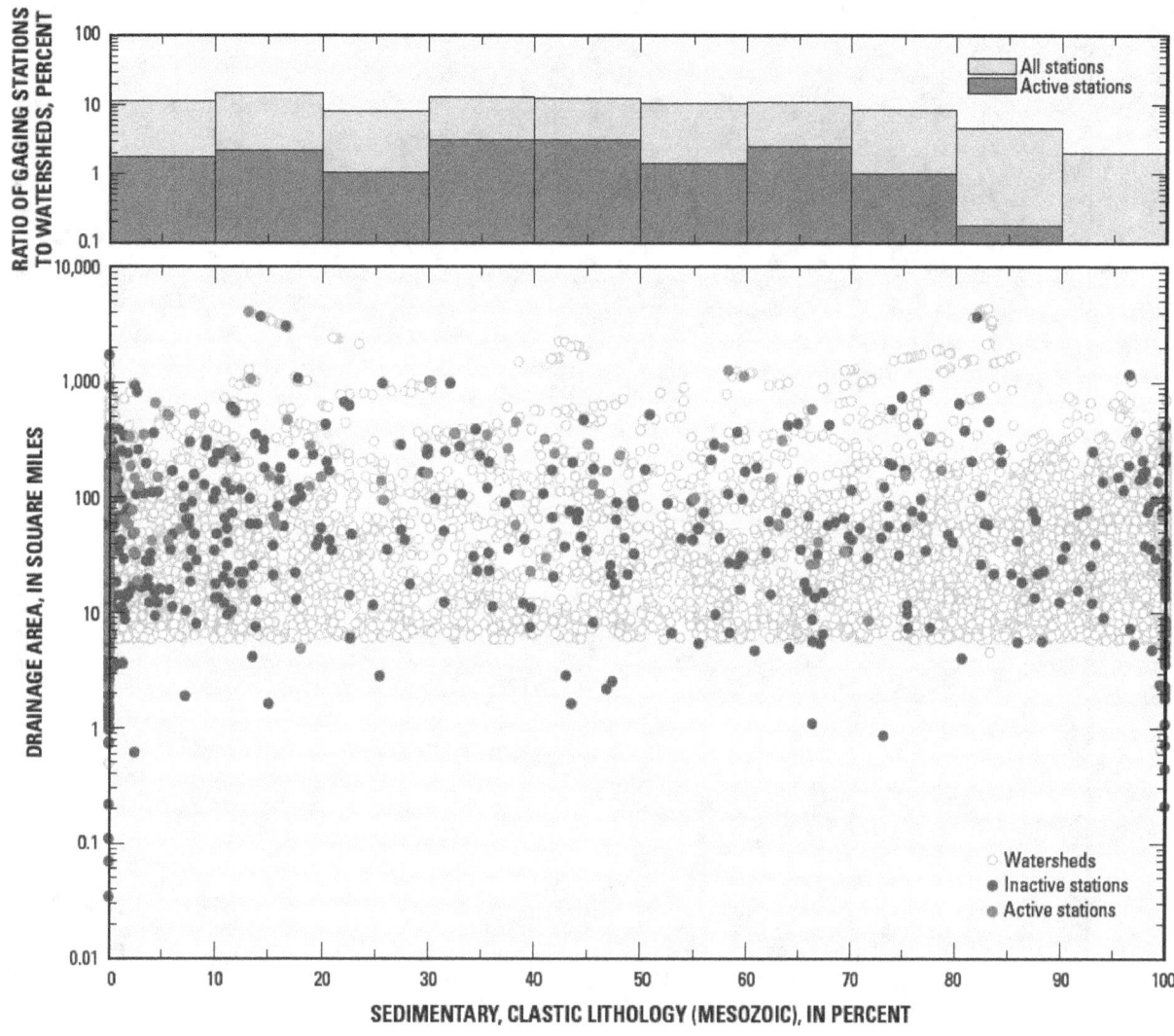

Figure 31.   Percentage of drainage basin area having sedimentary, clastic lithology (Mesozoic) for watersheds and U.S. Geological Survey streamgages in the Upper Colorado River Basin that are unaffected by reservoir regulation.

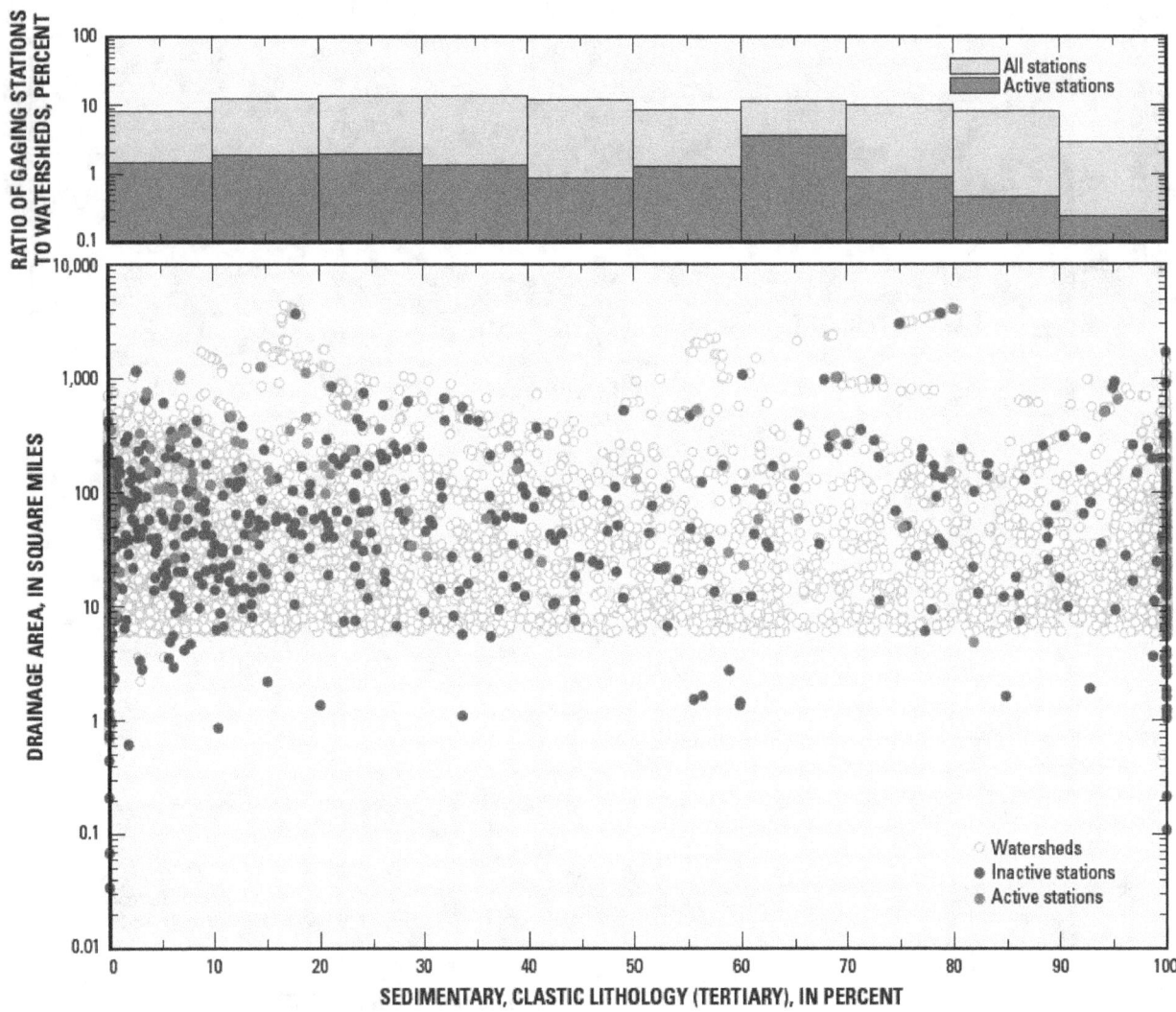

Figure 32.   Percentage of drainage basin area having sedimentary, clastic (continental) lithology (Tertiary) for watersheds and U.S. Geological Survey streamgages in the Upper Colorado River Basin that are unaffected by reservoir regulation.

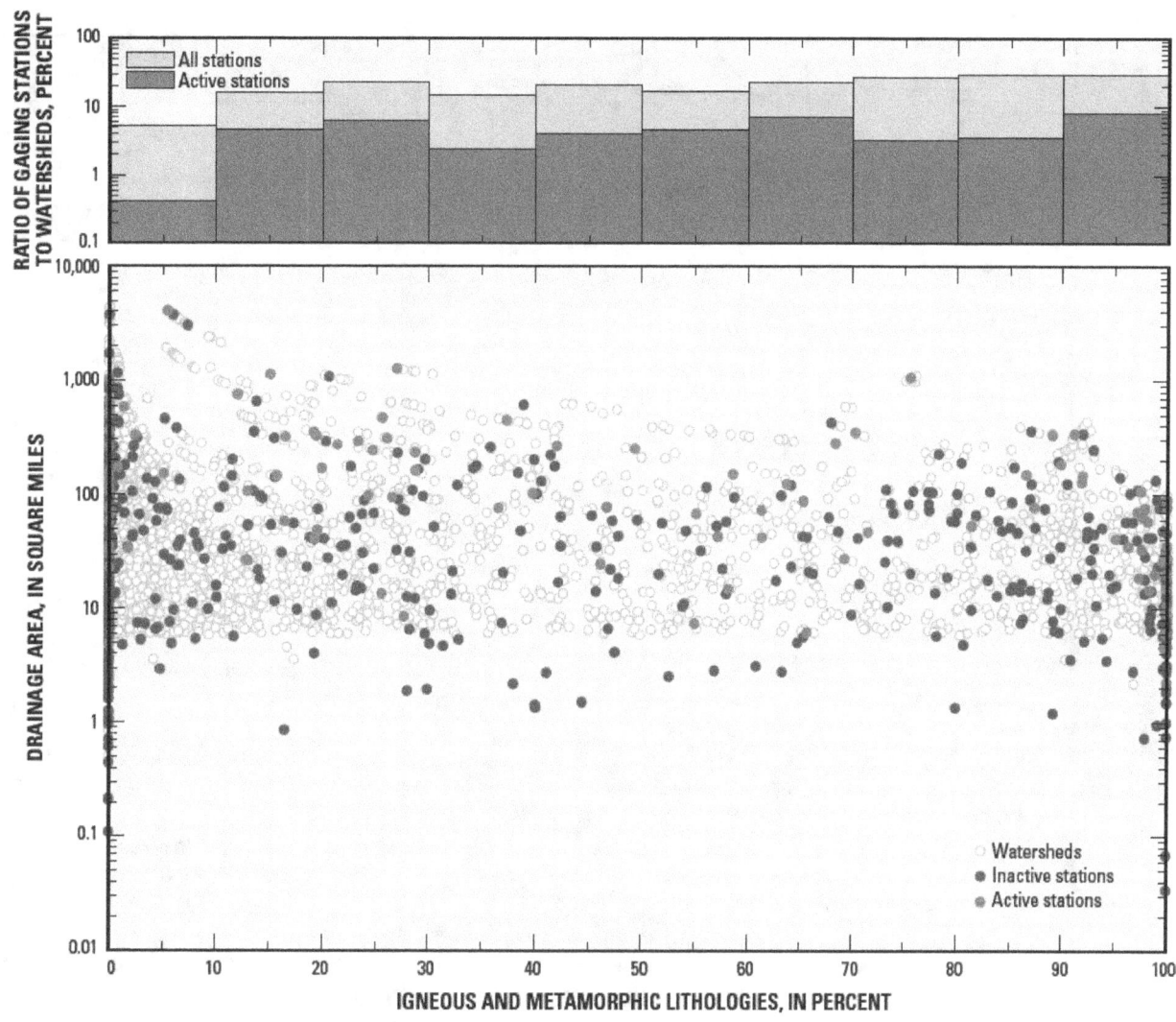

Figure 33.   Percentage of area having igneous or metamorphic lithologies for watersheds and U.S. Geological Survey streamgages in the Upper Colorado River Basin that are unaffected by reservoir regulation.

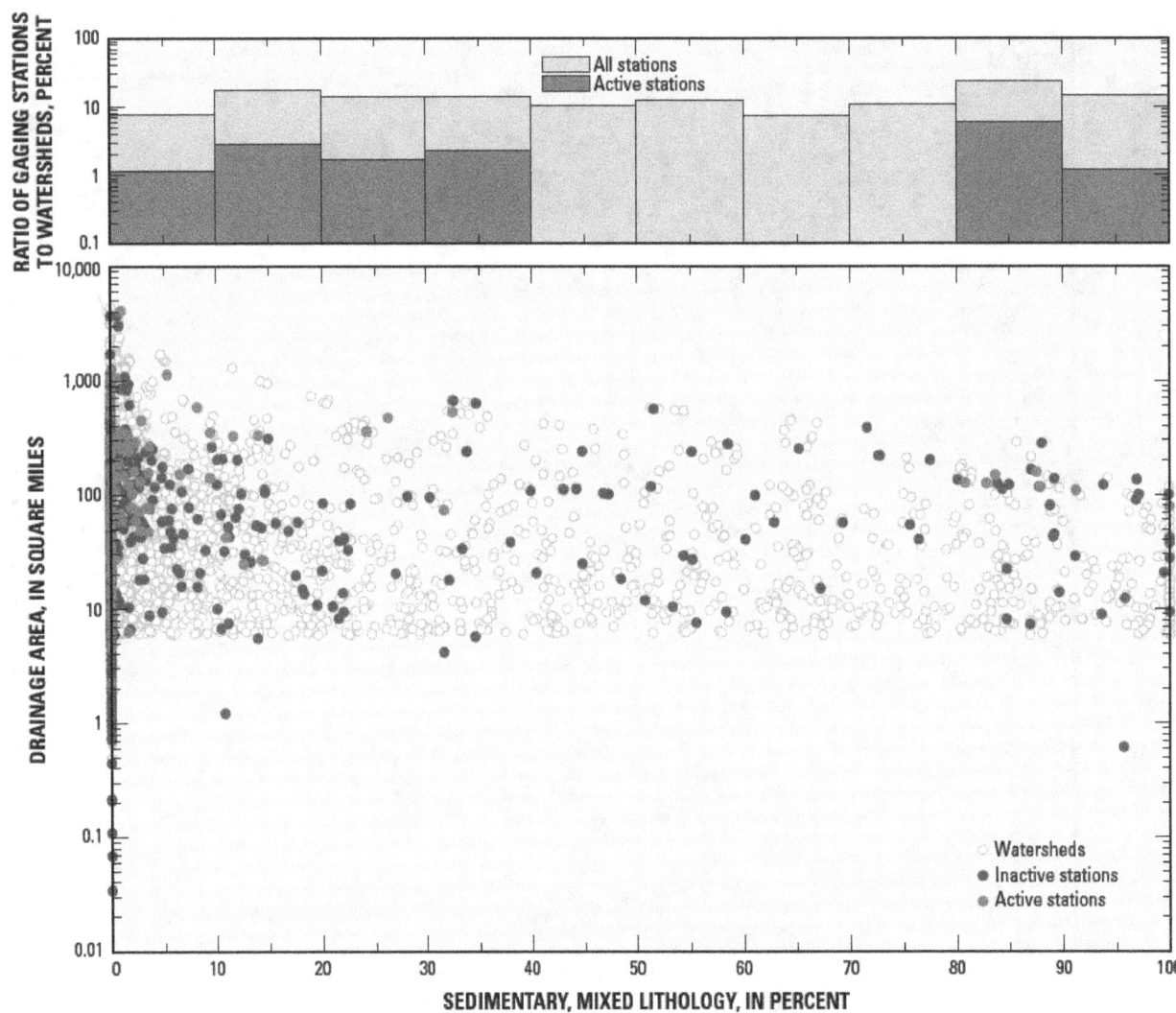

Figure 34. Percentage of drainage basin area having sedimentary, mixed (continental and marine) lithology for watersheds and U.S. Geological Survey streamgages in the Upper Colorado River Basin that are affected by reservoir regulation.

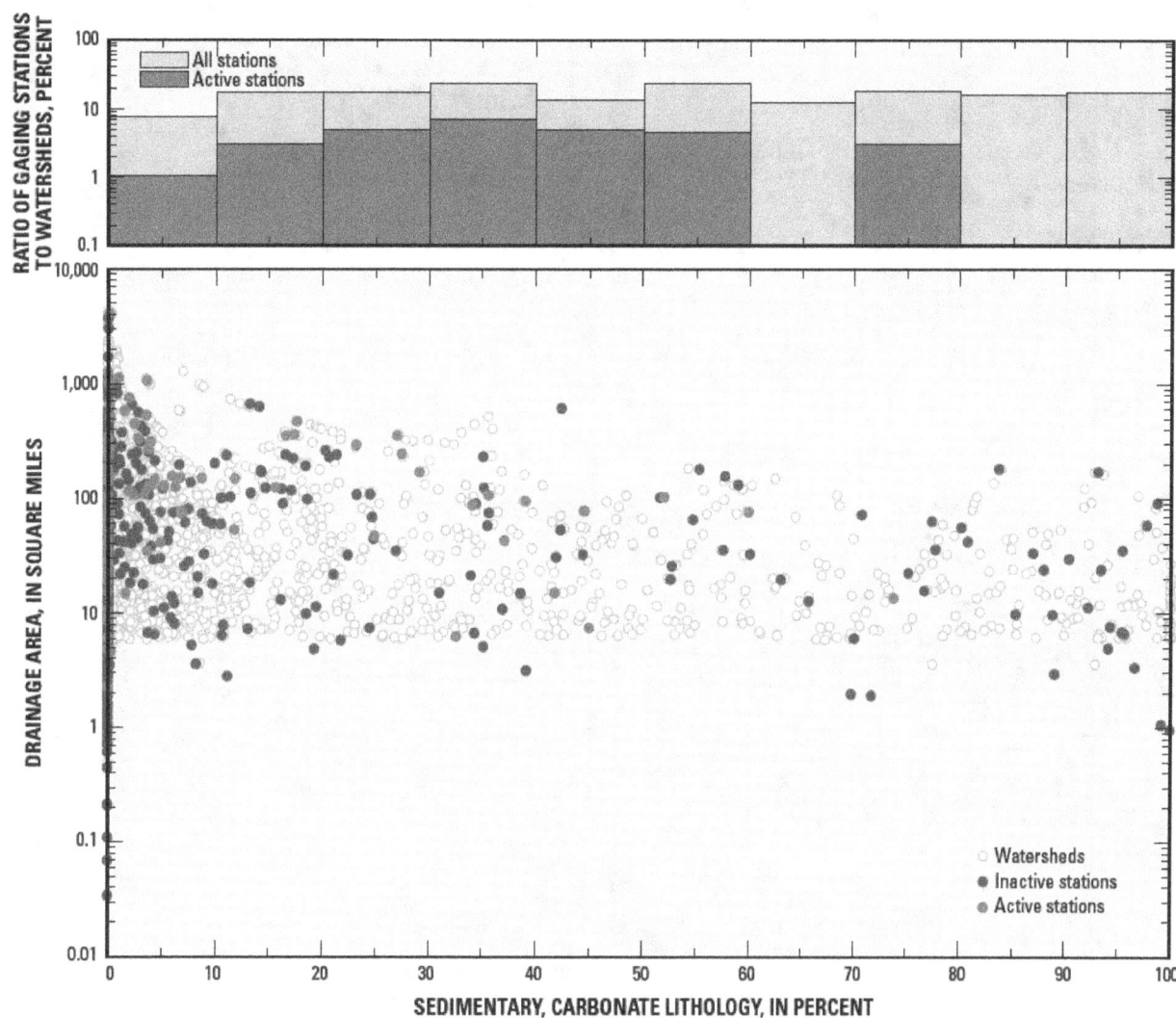

Figure 35.   Percentage of drainage basin area having sedimentary, carbonate (marine) lithology for watersheds and U.S. Geological Survey streamgages in the Upper Colorado River Basin that are unaffected by reservoir regulation.

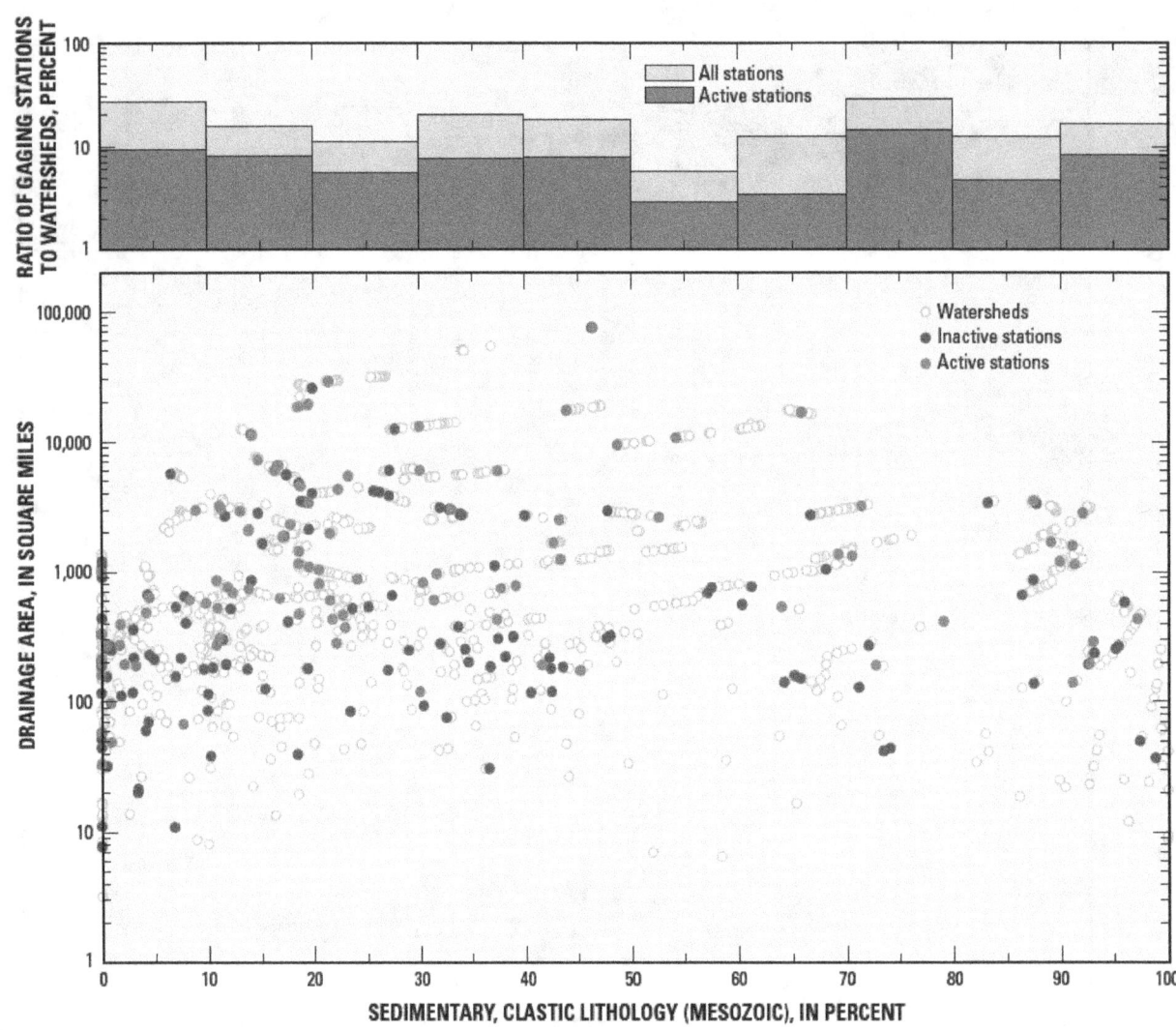

Figure 36.   Percentage of drainage basin area having sedimentary, clastic lithology (Mesozoic) for watersheds and U.S. Geological Survey streamgages in the Upper Colorado River Basin that are affected by reservoir regulation.

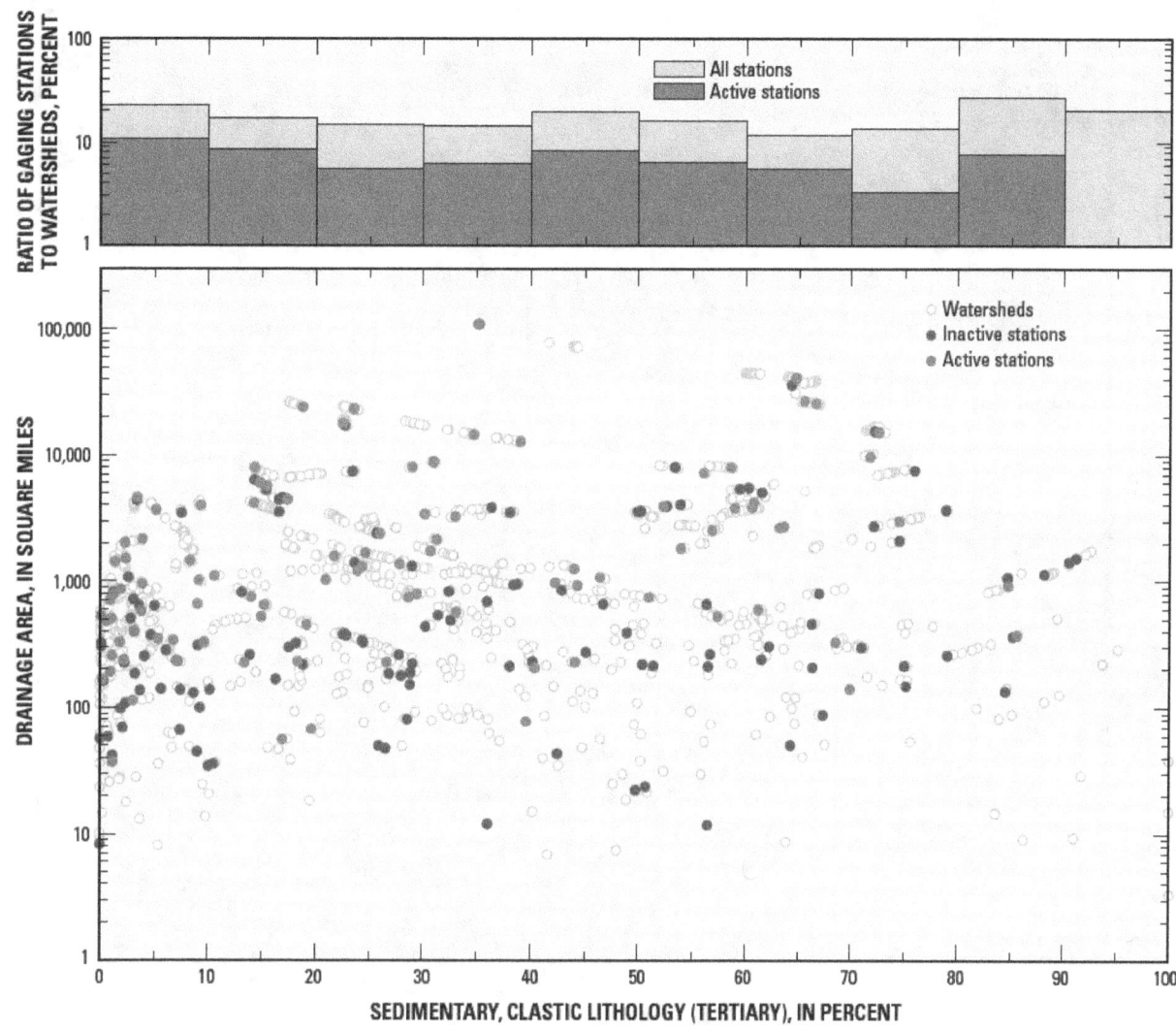

Figure 37.   Percentage of drainage basin area having sedimentary, clastic (continental) lithology (Tertiary) for watersheds and U.S. Geological Survey streamgages in the Upper Colorado River Basin that are affected by reservoir regulation.

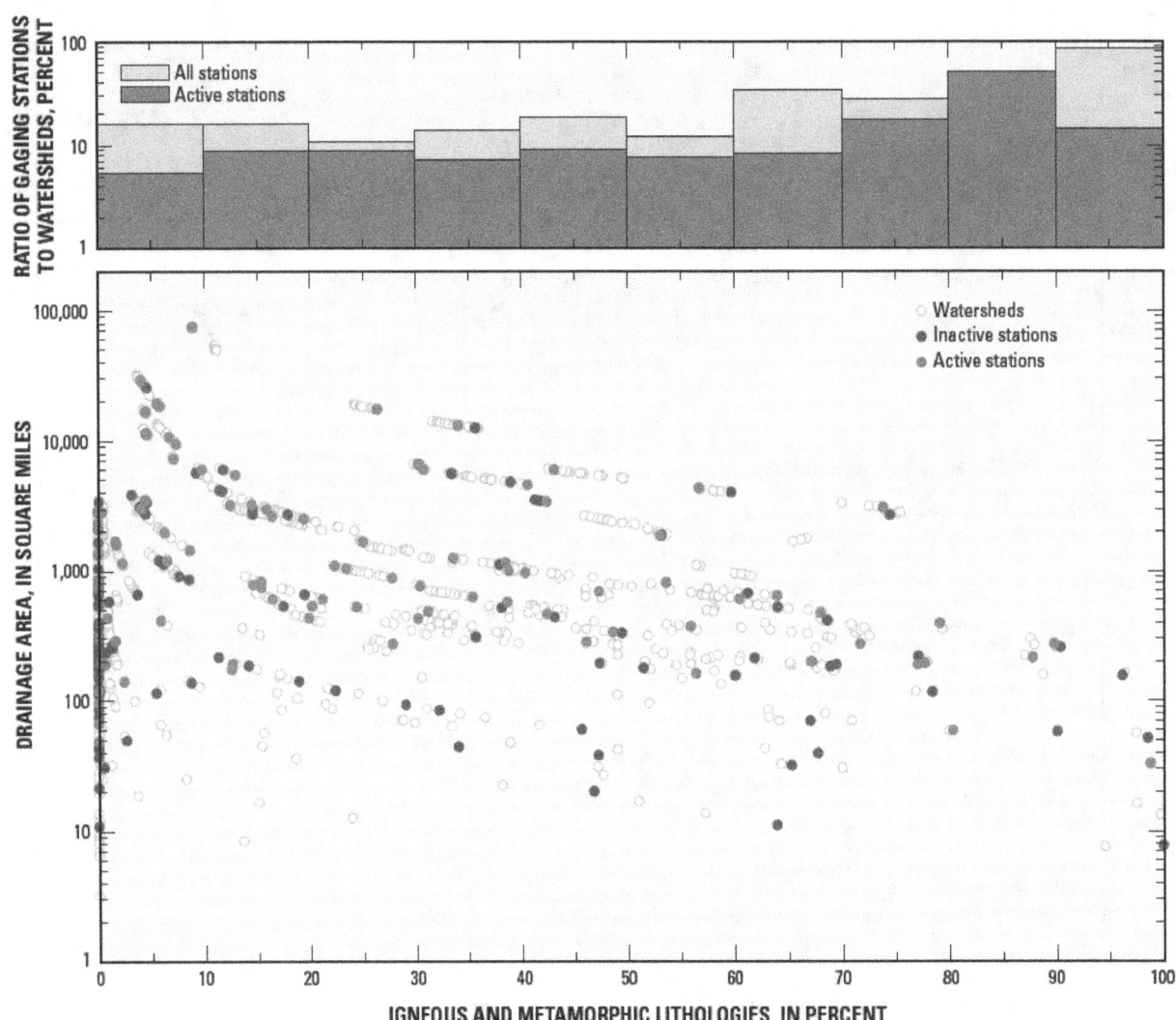

Figure 38.   Percentage of drainage basin area having igneous or metamorphic lithology for watersheds and U.S. Geological Survey streamgages in the Upper Colorado River Basin that are affected by reservoir regulation.

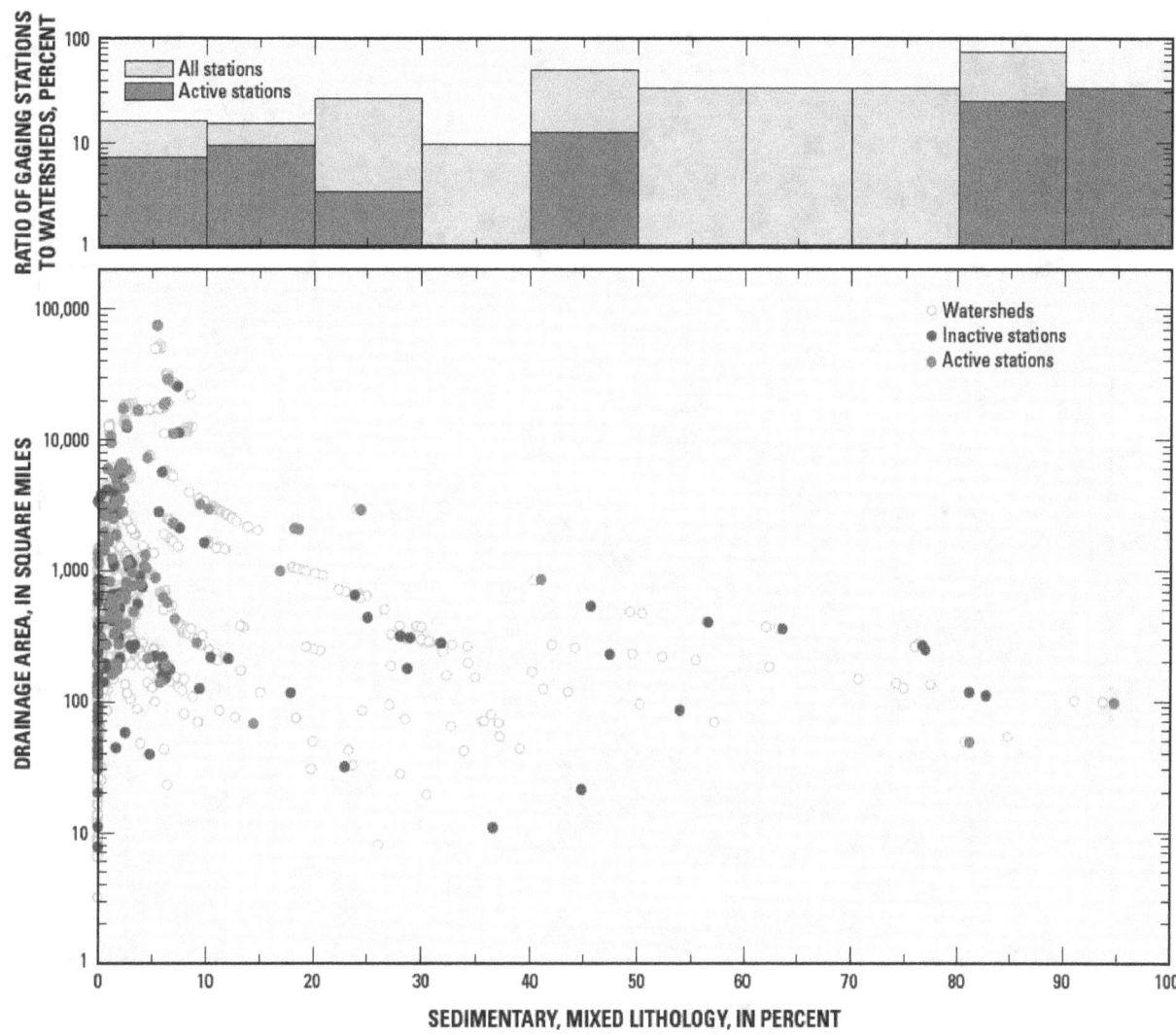

Figure 39.   Percentage of drainage basin area having sedimentary, mixed (continental and marine) lithology for watersheds and U.S. Geological Survey streamgages in the Upper Colorado River Basin that are affected by reservoir regulation.

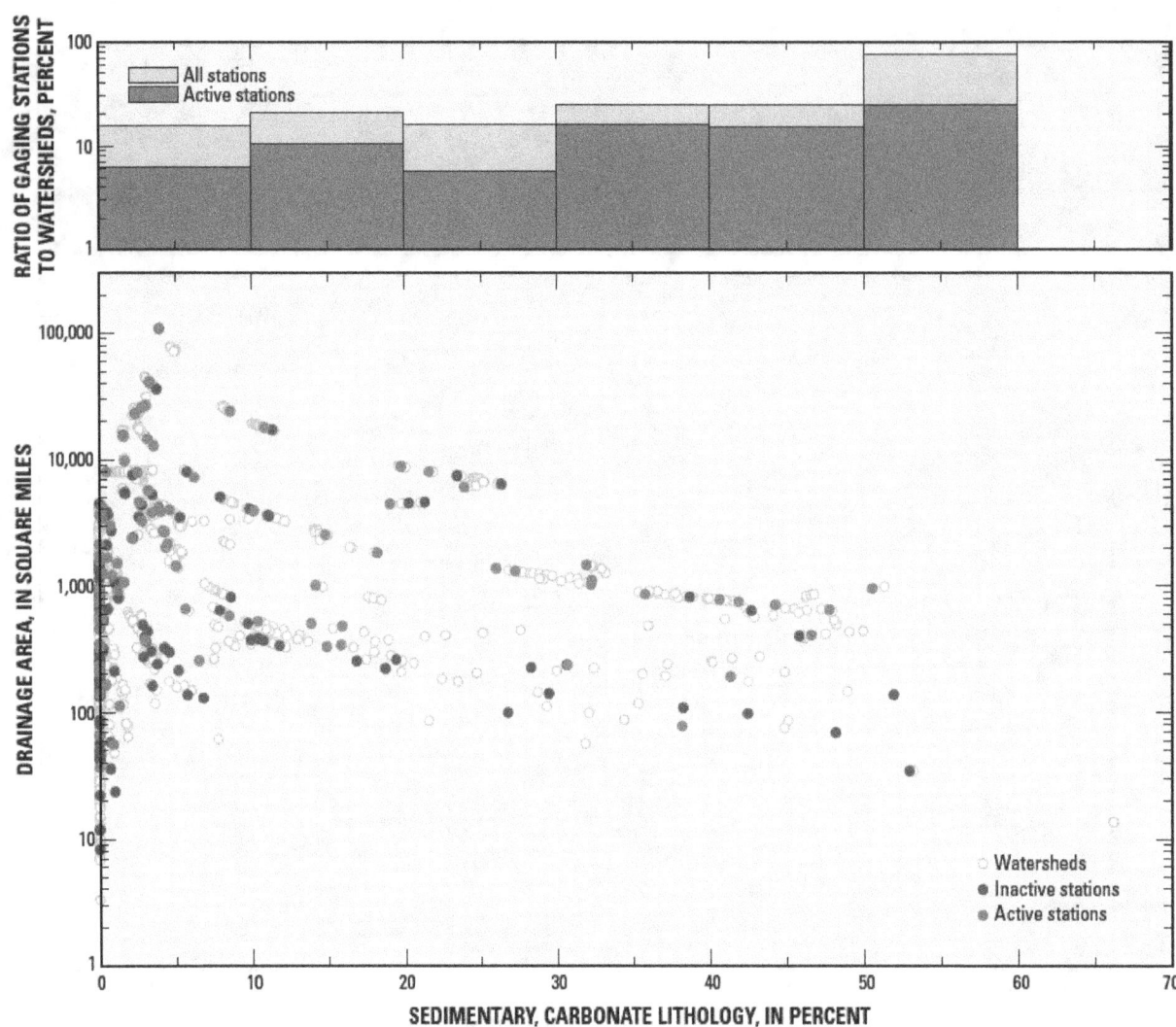

Figure 40.   Percentage of drainage basin area having sedimentary, carbonate (marine) lithology for watersheds and U.S. Geological Survey streamgages in the Upper Colorado River Basin that are affected by reservoir regulation.

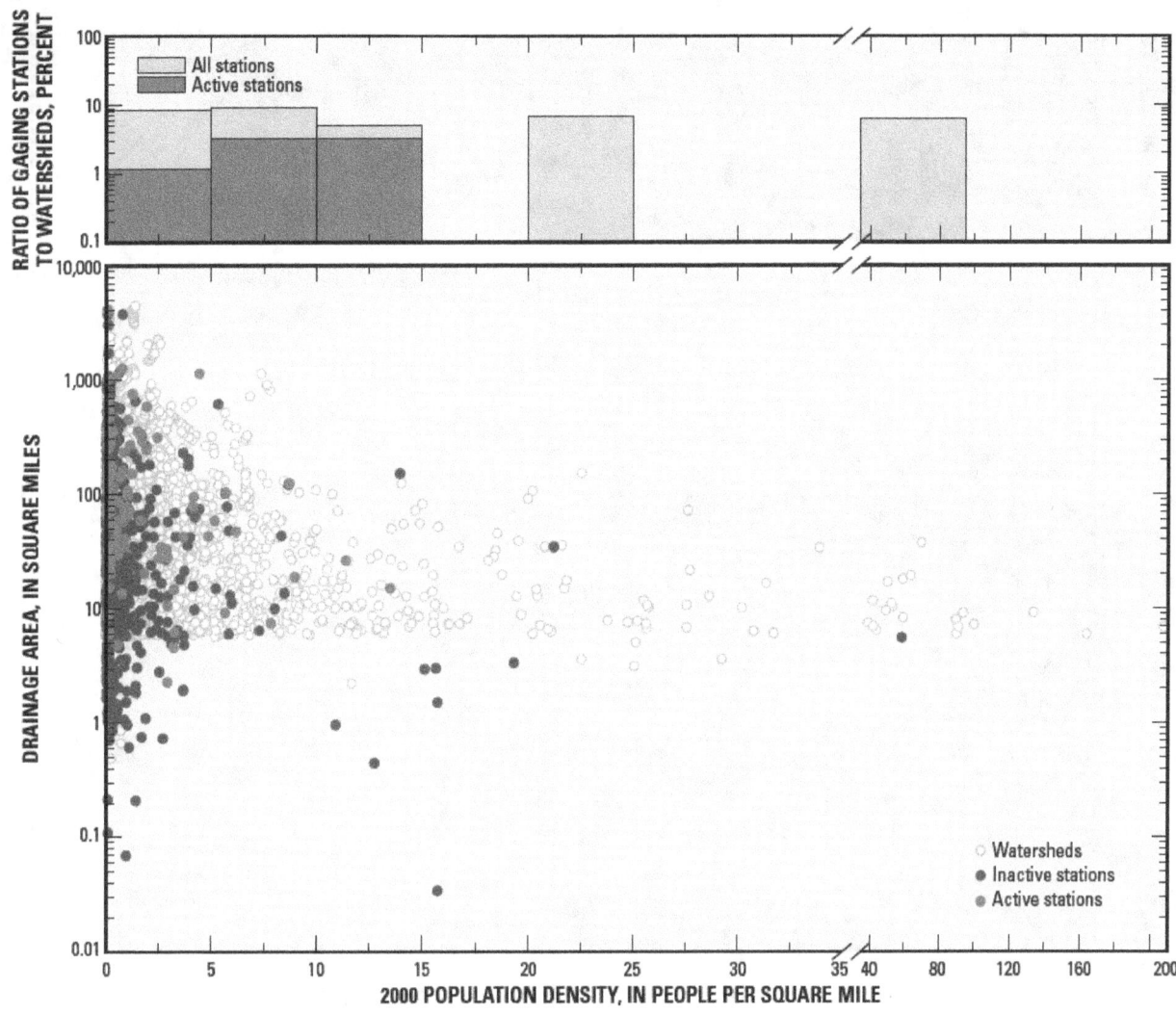

Figure 41.   Population density for watersheds and U.S. Geological Survey streamgage locations in the Upper Colorado River Basin that are unaffected by reservoir regulation.

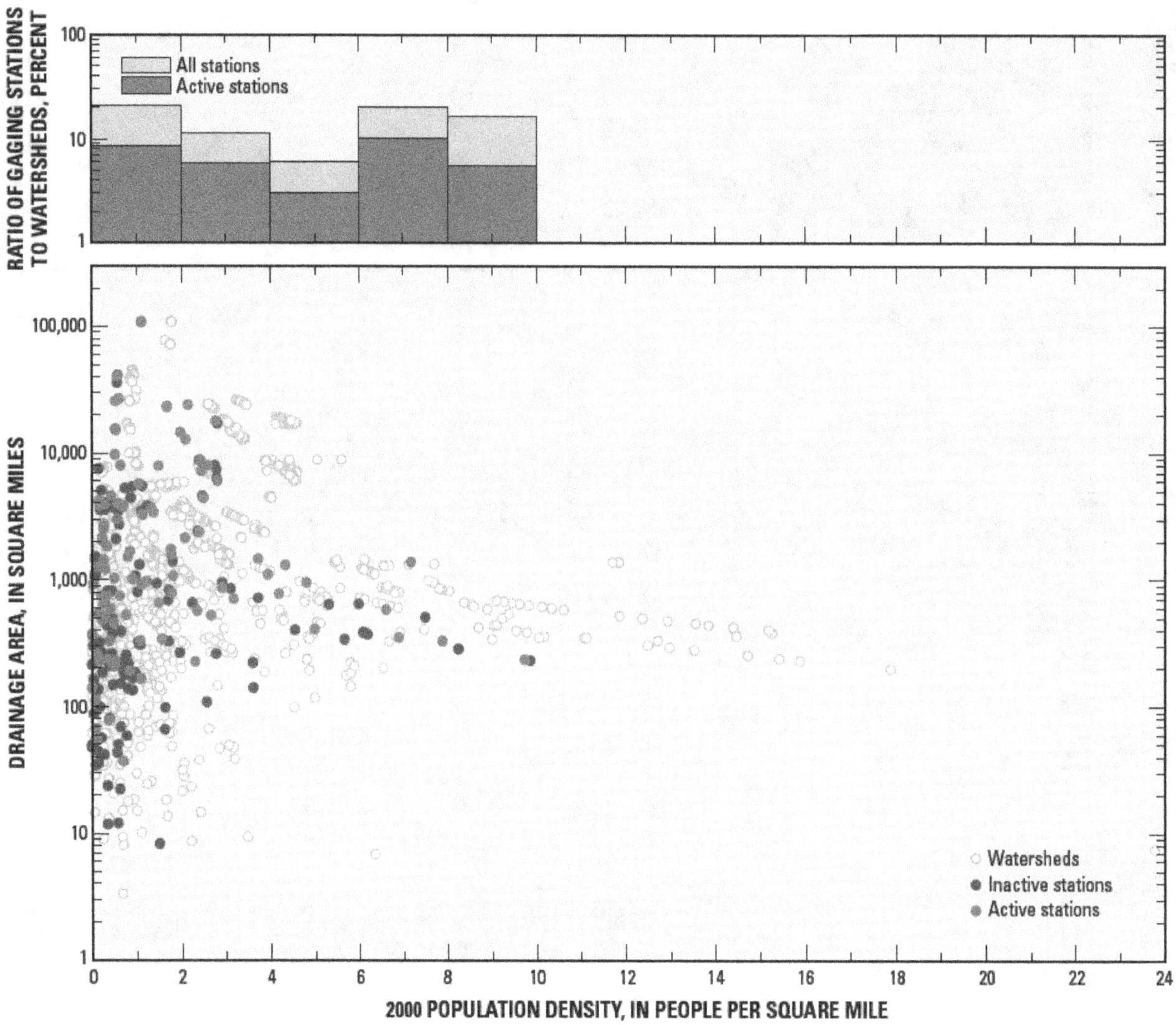

Figure 42.   Population density for watersheds and U.S. Geological Survey streamgage locations in the Upper Colorado River Basin that are affected by reservoir regulation.

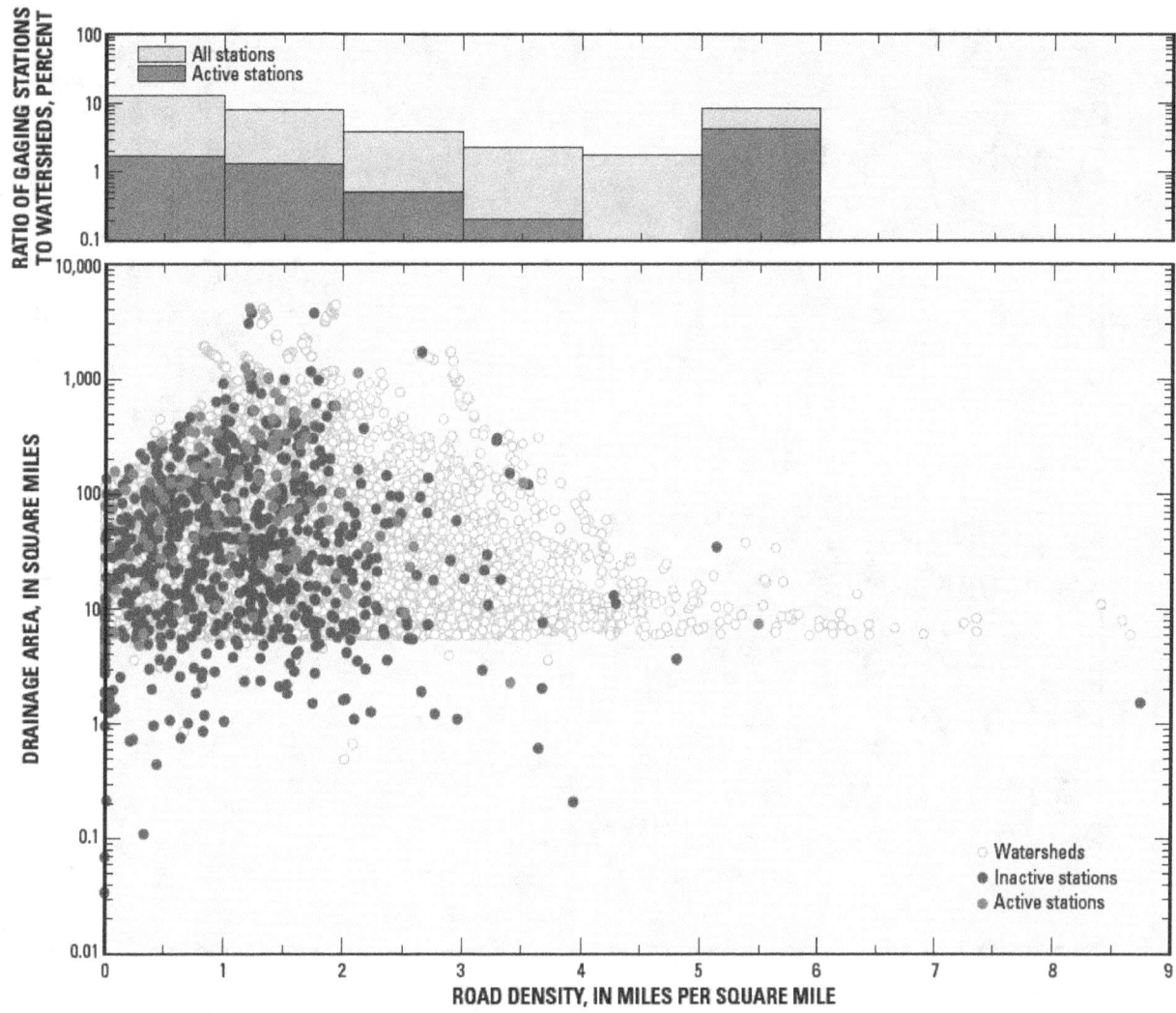

Figure 43.   Road density for watersheds and U.S. Geological Survey streamgage locations in the Upper Colorado River Basin that are unaffected by reservoir regulation.

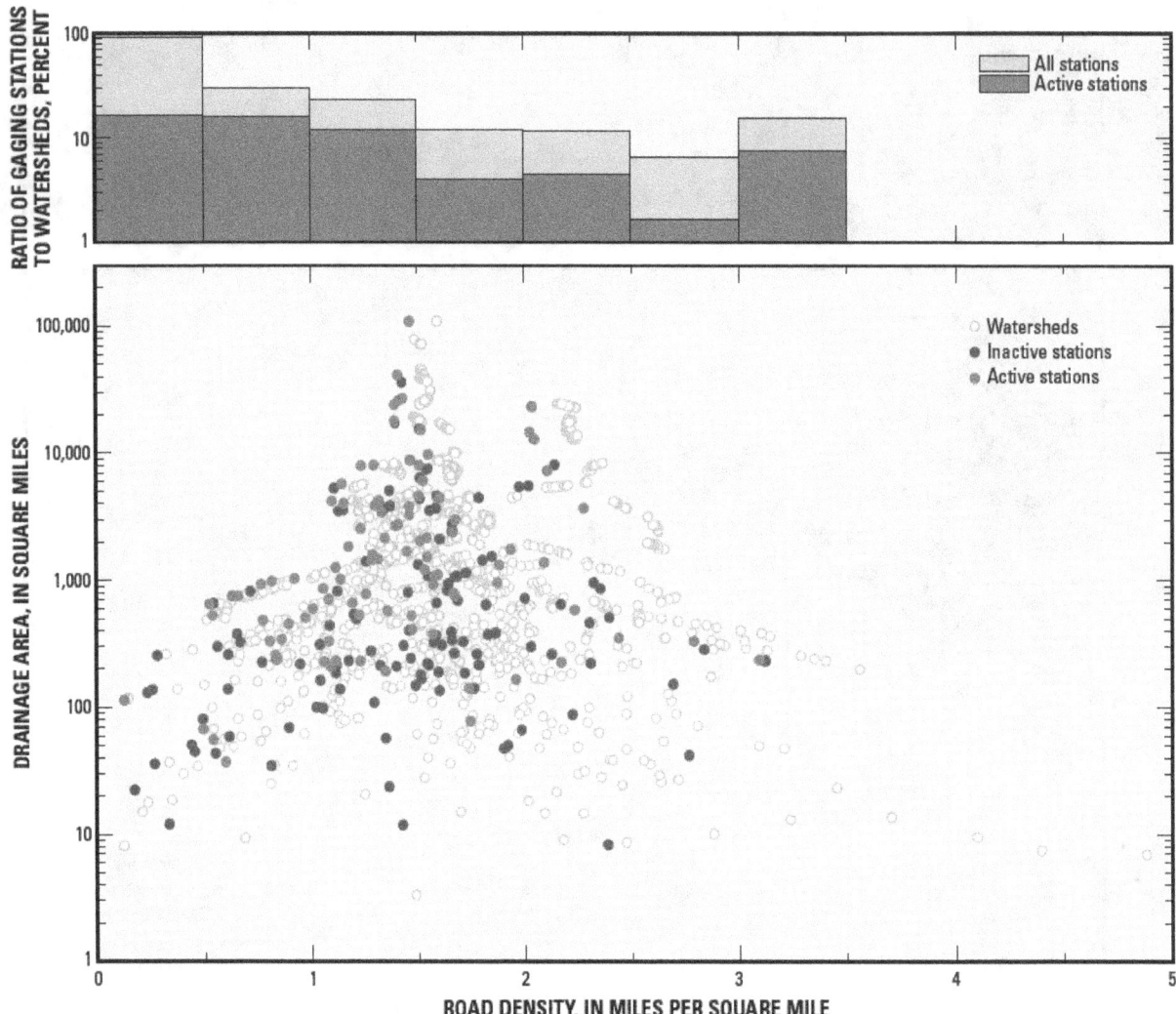

Figure 44.   Road density for watersheds and U.S. Geological Survey streamgage locations in the Upper Colorado River Basin that are affected by reservoir regulation.

Summary

The U.S. Geological Survey (USGS) has been gaging streamflow in the Upper Colorado River Basin (UCRB) since 1894. Since then, the USGS has operated 1,053 streamgages, of which 223 are active as of 2010. To assess how the USGS streamgage network represents the landscapes in the UCRB, 17 watershed characteristics were computed for each of the watersheds associated with these streamgages. The 17 watershed characteristics include physiographic parameters, land cover types, lithology, and parameters that describe anthropogenic influence. A set of 10,338 watersheds in the UCRB was constructed from a previously developed stream-reach network and the same 17 watershed characteristics were computed for each watershed. This set of watersheds was used to assess how well the streamgage network, both historically and currently, represents the landscapes in the UCRB. Computed watershed characteristics for UCRB watersheds and each

watershed instrumented with a USGS streamgage were plotted together to graphically identify watersheds, in terms of drainage area and the watershed characteristics, that are represented well by the USGS streamgage network and watersheds that are not represented well by the network. Active streamgages were identified in these plots to assess how the active network compares with the inactive network as well as with the watersheds in the basin. In many watersheds on the Colorado Plateau, much of the streamflow is ephemeral or intermittent. These types of watersheds are rarely monitored with streamgages, which creates an inherent bias in the USGS streamgage network toward watersheds that have perennial streamflow.

Watersheds were divided into those that are unaffected by upstream reservoir regulation and those affected by upstream reservoir regulation. In most cases, the streamgage network adequately represented the range of basin characteristics considered in this report for the unregulated watersheds of the UCRB, although the active USGS streamgage network tended to underrepresent most basin characteristics in unregulated watersheds. In regulated watersheds, the streamgage network, including the active network, generally represented the range of most basin characteristics in the watersheds of the UCRB.

References Cited

Buto, S.G., Kenney, T.A., and Gerner, S.J., 2010, Land disturbance associated with oil and gas development and effects of development-related land disturbance on dissolved-solids loads in streams in the Upper Colorado River Basin, 1991, 2007, and 2025: U.S. Geological Survey Scientific Investigations Report 2010–5064, 56 p. Available at http://pubs.usgs.gov/sir/2010/5064/.

Corbett, D.M., and others, 1943, Stream-gaging procedure; a manual describing methods and practices of the Geological Survey: U.S. Geological Survey Water-Supply Paper 888, 245 p.

Hitt, K.J., 2003, 2000 population density by block group for the conterminous United States: U.S. Geological Survey raster digital data, accessed August 2009 at http://water.usgs.gov/GIS/metadata/usgswrd/XML/uspopd00x10g.xml.

Homer, C., Huang, C., Yang, L., Wylie B., and Coan, M., 2004, Development of a 2001 national land cover database for the United States: Photogrammetric Engineering and Remote Sensing, v. 70, no. 7, p. 829–840, accessed April 2010 at http://www.mrlc.gov/publications.php.

Kenney, T.A., Gerner, S.J., Buto, S.G., and Spangler, L.E., 2009, Spatially referenced statistical assessment of dissolved-solids load sources and transport in streams of the Upper Colorado River Basin: U.S. Geological Survey Scientific Investigations Report 2009–5007, 50 p. Available at http://pubs.usgs.gov/sir/2009/5007.

PRISM Group, Oregon State University, 2007, Digital climate data, accessed August 2009 at http://www.ocs.oregonstate.edu/prism/index.phtml.

Ruddy, B.C. and Hitt, K.J., 1990, Summary of selected characteristics of large reservoirs: U.S. Geological Survey Open-File Report 90–163, accessed March 2010 at http://water.usgs.gov/lookup/getspatial?reservoir.

U.S. Geological Survey, 1999, The National Hydrography Dataset: U.S. Geological Survey Fact Sheet 109–99, accessed September 2006 at http://erg.usgs.gov/isb/pubs/factsheets/fs10699 html.

U.S. Geological Survey, 2010, The National Map: Transportation, vector digital data, accessed July 2010 at http://nationalmap.gov/transport html.

Appendix 1. U.S. Geological Survey streamflow-gaging stations in the Upper Colorado River Basin.

Appendix 1–1. Watershed characteristics for the U.S. Geological Survey streamgage network in the Upper Colorado River Basin.

Site number	Period of record	Latitude (decimal degrees)	Longitude (decimal degrees)	Drainage area (square miles)	Amount of upstream reservoir storage (acre-feet)	Mean basin elevation (feet)	Mean basin average annual precipitation (inches)	Area covered by developed land (percent)	Area covered by barren land (percent)
09010100	1969–75	40.465539	-105.846678	0.07	0	10,924	38.9	0.0	2.0
09010500	1953 to current year	40 325818	-105.856679	63.70	0	10,545	34.7	1.4	9.6
09010501	1953–67	40 325818	-105.856679	63.70	0	10,545	34.7	1.4	9.6
09011000	1904–18, 1933–86	40 218874	-105.857513	101.58	0	10,201	31.7	1.3	6.7
09011500	1950–55	40 252485	-105.834179	1.51	0	8,648	19.3	10.7	0.0
09012500	1905–09, 1910–12, 1947–55	40 253318	-105.811400	45.66	0	10,672	36.7	0.1	17.0
09013500	1947–55	40 236374	-105.798344	27.54	0	10,846	37.6	0.0	17.7
09014000	1904–09, 1910–13	40 244707	-105.826401	76.03	0	10,666	36.5	0.4	16.5
09015000	1947–59	40 206652	-105.838901	184.35	18,400	10,331	33.1	1.3	10.6
09015500	1950–55	40 186374	-105.820289	7.35	0	10,199	31.6	0.0	8.6
09016000	1951–55	40 130819	-105.767508	5.76	0	10,811	37.4	0.0	12.1
09016500	1944–71	40 112486	-105.749730	47.88	0	10,635	36.2	0.0	21.3
09018000	1950–55	40 188041	-105.895014	16.87	0	9,371	22.8	0.0	0.1
09019000	1950–82	40 144152	-105.867235	309.21	18,400	10,110	31.0	0.9	10.0
09019500	1907–11, 1953–33, 1961 to current year	40 120819	-105 900569	326.07	558,400	10,039	30.3	0.9	9.5
09020000	1934–53	40 180541	-106.009185	6.45	0	9,518	24.1	0.1	0.0
09020500	1953–60	40 155541	-105 980573	127.79	0	9,565	25.2	0.5	0.5
09021000	1953–82	40 145819	-105 940016	133.96	10,600	9,514	24.7	0.5	0.4
09022000	1908, 1909, 1968–73, 1984 to current year	39.845820	-105.751951	10.49	0	11,223	36.6	5.3	12.6
09023500	1907–09, 1934–37	39.880543	-105.756118	5.70	0	10,937	33.1	1.0	23.6
09024000	1910 to current year	39.899987	-105.776674	27.55	0	10,785	33.2	3.5	13.0
09025000	1907, 1909, 1933 to current year	39 920265	-105.785286	28.45	0	10,911	31.2	1.1	7.1
09025300	1996 to current year	39.889431	-105.832510	2.27	0	10,605	29.4	0.0	0.0
09025400	1970–96	39 919153	-105.825843	7.52	0	9,910	27.0	0.0	0.0
09026500	1933 to current year	39 909987	-105.878345	32.95	0	10,741	28.5	0.1	6.2
09032000	1934 to current year	39 949987	-105.765563	19.96	0	10,444	31.5	0.1	5.9
09032100	1983 to current year	39 985820	-105.745006	4.73	0	10,874	38.2	0.0	24.1
09032500	1934–60	39 997487	-105.823621	51.12	0	9,970	29.5	0.3	4.6
09033000	1935–56	40.050820	-105.777508	8.22	5,750	10,506	35.9	0.0	4.9
09033100	1997–2005	39 999153	-105.827510	65.67	5,750	9,966	29.7	0.3	4.2
09033300	1988 to current year	40.005820	-105.848344	223.94	5,750	10,050	28.0	1.8	4.7
09033500	1935–45	40.086653	-105.794731	3.04	0	9,949	30.4	0.0	0.0
09034000	1904–09, 1937–55	40.085264	-105 955294	259.45	5,750	9,902	27.0	1.9	4.0
09034250	1981 to current year	40 108319	-106.004185	787.56	574,750	9,770	27.1	1.3	5.4
09034500	1904–94	40.083319	-106.088076	823.64	574,750	9,734	26.7	1.3	5.1
09034800	1953–65	39 963041	-106.068909	6.63	0	9,384	22.1	0.0	0.0
09034900	1965 to current year	39.760264	-105 906401	6.03	0	11,793	31.5	0.0	29.5
09035500	1933–41, 1965 to current year	39.778875	-105 928347	16.50	0	11,556	29.6	0.0	20.6
09035700	1965 to current year	39.797208	-106.026129	35.27	0	11,107	28.1	0.0	11.9
09035800	1965 to current year	39.800542	-106.026407	8.85	0	10,937	28.0	0.0	5.6
09035820	1984–87	39.708320	-105 947514	2.75	0	11,742	32.9	0.0	17.1
09035830	1984–88	39.704153	-105 962515	4.06	0	11,635	32.2	0.0	18.6
09035840	1984–87	39.703598	-105 982238	5.51	0	11,517	31.5	0.0	18.7
09035845	1984–88	39.702764	-105 985572	6.16	0	11,493	31.3	0.0	20.1
09035850	1984–87	39.703042	-105 990294	6.50	0	11,464	31.2	0.0	19.2
09035870	1984–87	39.749153	-106.031962	20.13	0	11,193	29.5	0.0	12.2
09035880	1985–88	39.758875	-106.036129	21.78	0	11,152	29.3	0.0	11.7
09035900	1965 to current year	39.795820	-106.030573	27.47	0	10,984	28.7	0.0	9.5
09036000	1933 to current year	39.833875	-106.056408	89.47	0	10,890	28.0	0.1	8.8
09036500	1942–52	39 907486	-106.017240	13.85	0	10,531	28.5	0.0	5.9
09037000	1910–17	39 905264	-106.099465	140.11	0	10,487	26.5	0.3	6.5
09037200	1958–65	39 905264	-106 169468	2.80	0	9,891	23.0	0.2	1.0

Area covered by deciduous forest (percent)	Area covered by evergreen forest (percent)	Area covered by mixed forest (percent)	Area covered by shrubs young or stunted trees (percent)	Area covered by grass or herbaceous land (percent)	Sedimentary, clastic lithology, Mesozoic (percent)	Sedimentary, clastic (continental) lithology, Tertiary (percent)	Igneous or metamorphic lithologies (percent)	Sedimentary, mixed (continental and marine) lithology (percent)	Sedimentary, carbonate (marine) lithology (percent)	Population density (people per square mile)	Road density (miles per square mile)
0.0	68.5	0.0	0.0	29.4	0.0	0.0	100.0	0.0	0.0	0.97	0.00
0.1	58.0	0.0	0.0	8.7	1.4	0.7	97.7	0.0	0.0	0.96	0.93
0.1	58.0	0.0	0.0	8.7	1.4	0.7	97.7	0.0	0.0	0.96	0.93
0.2	65.7	0.0	0.1	7.4	0.9	2.5	96.5	0.0	0.0	0.96	1.17
0.6	79.2	0.0	1.1	1.8	0.0	55.5	44.5	0.0	0.0	0.97	8.75
0.0	52.2	0.0	0.0	10.6	0.0	0.0	100.0	0.0	0.0	0.96	0.55
0.1	46.9	0.0	0.3	8.2	0.0	0.0	100.0	0.0	0.0	0.96	0.06
0.0	50.8	0.0	0.2	9.2	0.0	0.7	99.3	0.0	0.0	0.96	0.50
0.1	59.0	0.0	0.2	7.9	0.5	3.3	96.2	0.0	0.0	0.96	1.11
0.2	79.2	0.0	0.1	3.8	0.0	0.4	99.6	0.0	0.0	0.97	0.33
0.1	42.9	0.0	0.0	8.6	0.0	0.3	99.7	0.0	0.0	0.97	0.00
0.4	48.0	0.0	0.0	8.7	0.0	0.4	99.6	0.0	0.0	1.27	0.34
3.0	76.1	0.0	3.2	8.4	3.8	54.0	42.2	0.0	0.0	1.23	1.38
0.4	59.2	0.0	1.1	7.3	0.5	9.1	90.4	0.0	0.0	1.05	1.04
0.6	59.6	0.0	1.8	7.1	0.5	9.7	89.8	0.0	0.0	1.08	1.08
5.6	78.1	0.0	5.4	8.5	0.2	99.8	0.0	0.0	0.0	1.50	1.77
1.4	85.1	0.0	2.2	7.1	9.0	86.7	4.3	0.0	0.0	0.87	1.66
1.4	83.2	0.0	4.0	6.9	10.0	84.6	5.4	0.0	0.0	0.90	1.61
0.0	43.9	0.0	0.0	20.3	0.0	0.0	99.9	0.0	0.0	2.93	0.94
1.9	48.6	0.0	0.0	10.0	0.0	0.0	99.9	0.0	0.0	2.94	0.73
0.4	55.1	0.0	0.0	15.1	0.0	0.0	99.9	0.0	0.0	2.93	1.62
0.1	63.5	0.0	0.0	14.0	0.0	0.0	99.7	0.3	0.0	2.93	1.06
0.0	90.0	0.0	0.0	3.5	0.0	0.0	100.0	0.0	0.0	2.93	3.41
0.3	94.2	0.0	0.1	1.9	0.0	1.7	98.3	0.0	0.0	2.93	3.68
0.0	73.3	0.0	0.0	6.2	2.4	0.0	97.6	0.0	0.0	2.89	1.17
0.3	71.2	0.0	0.0	11.9	0.0	1.1	98.9	0.0	0.0	1.82	1.96
0.0	48.2	0.0	0.0	12.4	0.0	0.0	100.0	0.0	0.0	1.50	0.67
0.7	75.2	0.0	1.2	8.6	0.0	9.3	87.5	3.2	0.0	1.65	2.00
0.0	76.7	0.0	0.0	6.1	0.0	0.0	100.0	0.0	0.0	1.50	2.39
0.6	77.4	0.0	1.1	7.6	0.0	7.4	90.1	2.5	0.0	1.62	1.99
0.8	70.3	0.0	1.6	9.6	3.2	13.4	77.0	6.3	0.0	2.30	2.17
0.1	96.8	0.0	0.0	1.8	0.0	0.0	100.0	0.0	0.0	1.51	0.75
0.9	70.2	0.0	3.6	8.9	3.0	13.9	77.0	6.0	0.0	2.77	2.13
1.2	64.3	0.0	6.2	7.8	4.5	29.5	63.9	2.0	0.0	1.78	1.67
1.3	64.0	0.0	7.0	7.9	4.3	32.5	61.2	2.0	0.0	1.72	1.64
6.7	80.1	0.2	2.7	8.5	0.0	53.1	46.9	0.0	0.0	0.15	1.28
0.0	24.2	0.0	0.0	8.2	0.0	0.0	99.9	0.0	0.0	0.17	0.32
0.0	33.8	0.0	0.0	10.0	0.0	0.0	100.0	0.0	0.0	0.17	0.41
0.6	54.8	0.0	0.0	8.3	0.0	0.0	100.0	0.0	0.0	0.18	0.37
0.3	69.1	0.0	0.0	8.4	0.0	0.0	100.0	0.0	0.0	0.20	0.22
0.0	15.4	0.0	0.0	8.5	0.0	0.0	100.0	0.0	0.0	0.18	0.00
0.0	19.3	0.0	0.0	12.4	0.0	0.0	100.0	0.0	0.0	0.18	0.00
0.0	26.6	0.0	0.0	12.6	0.0	0.0	100.0	0.0	0.0	0.17	0.26
0.0	26.5	0.0	0.0	13.0	0.0	0.0	100.0	0.0	0.0	0.17	0.27
0.0	28.8	0.0	0.0	13.1	0.0	0.0	100.0	0.0	0.0	0.17	0.30
0.2	48.4	0.0	0.0	13.1	0.0	0.0	100.0	0.0	0.0	0.17	0.36
0.3	50.1	0.0	0.0	12.7	0.0	0.0	100.0	0.0	0.0	0.17	0.40
0.2	57.8	0.0	0.0	11.1	0.0	0.0	100.0	0.0	0.0	0.17	0.44
0.5	62.3	0.0	0.0	8.4	0.0	0.0	100.0	0.0	0.0	0.17	0.50
0.0	81.1	0.0	0.0	5.5	1.1	10.1	88.8	0.0	0.0	0.23	1.40
0.5	68.6	0.0	0.3	9.1	0.9	3.6	95.5	0.0	0.0	0.17	0.82
10.9	79.8	1.6	0.5	1.4	0.0	3.2	96.8	0.0	0.0	0.16	0.93

Site number	Period of record	Latitude (decimal degrees)	Longitude (decimal degrees)	Drainage area (square miles)	Amount of upstream reservoir storage (acre-feet)	Mean basin elevation (feet)	Mean basin average annual precipitation (inches)	Area covered by developed land (percent)	Area covered by barren land (percent)
09037500	1904–24, 1933 to current year	40.000263	-106 179746	183.85	0	10,127	24.6	0.3	4.9
09038500	1948–54, 1958 to current year	40.035263	-106 205303	229.73	93,637	9,815	23.0	0.2	3.9
09039000	1953–93	40 217484	-106 313085	46.17	0	9,803	27.9	0.0	1.3
09039500	1937–43	40 208318	-106 309752	51.38	0	9,708	27.1	0.0	1.2
09040000	1937–43, 1953–83	40 157485	-106 283362	75.99	0	9,264	23.1	0.0	0.5
09040500	1904–05, 1921–22, 1937–56	40.059152	-106 305584	168.04	0	9,084	22.2	0.0	0.6
09041000	1937–43, 1955–71, 1993–99	40 293595	-106.483647	87.74	0	8,800	26.7	0.8	0.0
09041090	1990 to current year	40 202484	-106.422534	144.76	0	8,682	23.7	0.9	0.1
09041100	1955–68	40 240540	-106 373643	11.48	0	8,896	20.7	0.0	0.0
09041200	1955–74	40 161374	-106 559482	17.85	0	9,670	34.9	0.0	0.0
09041300	1957–70	40 132763	-106.490313	17.62	0	9,045	25.1	1.4	0.1
09041400	1995 to current year	40 108596	-106.413921	270.38	0	8,602	22.0	0.8	0.1
09041500	1904–05, 1982–95	40.060263	-106 398087	288.85	0	8,556	21.4	0.9	0.1
09043000	1953–58	39 381655	-106.063075	5.73	0	12,304	32.7	0.9	50.6
09044000	1953–58	39 373044	-106.054463	1.95	0	12,075	32.5	0.0	24.8
09044500	1953–58	39 373877	-106.048352	1.90	0	12,099	32.6	0.0	25.5
09045000	1953–58	39.403599	-106.051963	4.81	0	12,124	33.2	0.1	42.3
09045500	1953–58	39.440821	-106.041963	6.37	0	11,996	34.0	0.1	35.2
09046490	1983 to current year	39.455821	-106.031685	42.37	0	11,634	30.9	1.1	24.3
09046530	1955–2005	39.493042	-106.044741	11.01	0	11,032	28.0	3.2	20.0
09046600	1957 to current year	39 566653	-106.049463	123.19	0	10,978	27.4	3.1	13.1
09046601	1957–67	39 548598	-106.039185	119.04	0	11,024	27.7	3.1	13.6
09047500	1942–46, 1951 to current year	39.605542	-105.943070	57.72	0	11,494	29.9	1.1	19.1
09047700	1957 to current year	39 594431	-105 972516	9.09	0	10,869	26.6	0.0	1.6
09049200	1973–79	39 501098	-106 168356	22.16	0	11,158	28.9	2.4	4.9
09050000	1942–50	39 575264	-106 110576	86.58	0	11,206	29.2	2.5	14.0
09050100	1957 to current year	39 575264	-106 110576	86.58	0	11,206	29.2	2.5	14.0
09050700	1960 to current year	39.625542	-106.066408	328.42	252,678	10,954	27.2	3.4	12.3
09051000	1943–52	39.648598	-106.020018	12.81	0	11,206	29.1	4.0	8.4
09051050	1986 to current year	39.639709	-106.040296	18.63	0	11,057	27.8	4.0	5.9
09051500	1942–51	39.649986	-106.080298	13.51	0	10,860	27.1	0.7	21.0
09052000	1942–56, 1966–94	39.723042	-106 128633	15.84	0	10,737	28.8	0.0	20.9
09052400	1966–94	39.728041	-106 173357	8.70	0	11,169	32.0	0.0	27.4
09052500	1942–51	39.737208	-106.135855	9.91	0	10,949	30.9	0.0	24.0
09052800	1966–94	39.763041	-106.192524	14.12	0	11,032	32.5	0.0	30.1
09053000	1942–54	39.781652	-106.167801	16.60	0	10,735	30.8	0.0	25.6
09053500	1943–71, 1985–88	39.831930	-106 222803	504.37	252,678	10,643	26.3	3.1	11.1
09054000	1942–49, 1966–94	39.799152	-106 268360	15.21	0	11,163	33.9	0.0	30.9
09054500	1944–53	39.855819	-106 252527	18.38	0	10,756	31.5	0.0	25.6
09055000	1944–53	39.852763	-106 267805	7.47	0	9,606	24.2	0.0	7.6
09055300	1966–94	39.835263	-106 316417	12.03	0	10,815	31.6	0.0	18.1
09055500	1944–53	39.849986	-106 291416	15.55	0	10,579	30.1	0.0	15.0
09057500	1937 to current year	39.880264	-106 333918	576.81	406,678	10,529	26.0	2.8	11.0
09057520	1989–94	39 963597	-106 360307	638.59	406,678	10,391	25.5	2.6	10.0
09058000	1904–18, 1961 to current year	40.036652	-106.440032	2,372.28	1,075,065	9,604	24.1	1.4	4.9
09058030	1981–90	39 966931	-106 523367	2,402.68	1,075,065	9,588	24.0	1.4	4.8
09058500	1948–54, 1964–2004	39.708042	-106.426697	12.84	0	10,815	30.3	0.0	20.5
09058600	1964–71	39.703875	-106.457531	3.32	0	10,029	28.6	0.0	0.0
09058610	1972–2004	39.703875	-106.457531	3.32	0	10,029	28.6	0.0	0.0
09058700	1965–2004	39.698319	-106.445586	2.92	0	9,844	27.5	0.0	0.0
09058800	1965–2004	39.731653	-106.426697	3.56	0	10,575	29.6	0.0	3.9
09059500	1944 to current year	39.799987	-106 583923	93.37	0	9,718	25.8	0.0	3.7
09060500	1952–81	40.041098	-106.655872	47.91	0	9,392	29.5	0.5	0.1
09060550	1984–99	39 978321	-106.710039	72.73	0	9,222	27.7	0.3	0.1

Area covered by deciduous forest (percent)	Area covered by evergreen forest (percent)	Area covered by mixed forest (percent)	Area covered by shrubs young or stunted trees (percent)	Area covered by grass or herbaceous land (percent)	Sedimentary, clastic lithology, Mesozoic (percent)	Sedimentary, clastic (continental) lithology, Tertiary (percent)	Igneous or metamorphic lithologies (percent)	Sedimentary, mixed (continental and marine) lithology (percent)	Sedimentary, carbonate (marine) lithology (percent)	Population density (people per square mile)	Road density (miles per square mile)
1.7	65.3	0.1	5.7	8.2	2.0	7.2	89.9	0.8	0.0	0.17	0.93
2.6	58.9	0.2	13.0	7.3	2.2	18.4	77.7	1.7	0.0	0.17	1.12
3.0	77.9	0.2	6.0	9.1	0.0	14.1	85.8	0.0	0.0	0.15	0.84
3.2	75.6	0.2	7.8	8.6	0.0	14.9	85.1	0.0	0.0	0.15	0.84
2.8	79.5	0.1	6.8	7.8	0.0	89.6	10.4	0.0	0.0	0.15	0.37
2.4	60.4	0.1	20.6	8.0	0.7	63.0	34.2	2.1	0.0	0.15	0.83
13.2	34.6	0.3	23.8	17.3	52.6	23.4	23.9	0.0	0.0	0.19	1.46
12.5	26.3	0.3	35.9	15.5	62.3	21.1	15.7	0.8	0.0	0.17	1.44
13.3	22.9	0.1	42.7	17.7	24.8	59.6	15.5	0.0	0.0	0.15	0.96
4.6	80.1	0.4	0.1	8.7	12.6	4.4	83.0	0.0	0.0	0.18	0.92
9.9	63.4	2.0	15.8	4.7	28.3	5.6	62.8	0.0	3.3	0.15	2.76
9.8	26.8	0.4	41.9	12.0	57.8	18.9	21.4	1.6	0.2	0.17	1.44
9.3	25.3	0.4	44.3	11.8	56.9	20.7	20.3	1.8	0.2	0.43	1.52
0.1	11.0	0.0	0.0	21.8	0.0	0.0	78.2	0.0	21.8	3.72	1.78
0.0	25.8	0.0	0.0	46.8	0.0	0.0	29.9	0.0	69.7	3.69	0.07
0.0	23.8	0.0	0.0	48.1	0.0	0.0	28.0	0.0	71.6	3.69	0.00
1.6	18.0	0.0	0.0	23.3	0.0	0.0	80.7	0.0	19.3	3.73	1.34
0.2	29.4	0.0	0.0	23.6	0.0	0.0	89.3	0.0	10.7	7.30	2.02
1.1	40.6	0.0	0.0	24.4	5.6	0.0	57.2	0.0	37.1	4.87	2.32
2.6	59.9	0.0	0.0	11.3	39.8	0.0	54.0	0.8	5.2	5.98	4.29
2.4	56.3	0.1	0.1	19.7	18.9	0.0	63.8	1.6	15.6	8.73	3.51
2.2	55.8	0.1	0.0	20.0	17.9	0.0	64.2	1.7	16.2	8.70	3.55
1.7	39.4	0.0	0.0	31.4	1.8	0.0	98.2	0.0	0.0	5.17	1.46
0.3	81.3	0.1	0.0	15.9	0.0	0.0	100.0	0.0	0.0	5.68	2.52
1.7	46.8	0.0	0.0	37.4	0.0	0.0	24.9	0.0	75.1	0.35	1.25
2.4	43.7	0.1	0.0	30.9	0.0	0.5	65.4	0.0	34.2	0.56	1.73
2.4	43.7	0.1	0.0	30.9	0.0	0.5	65.4	0.0	34.2	0.56	1.73
2.8	50.5	0.2	0.3	23.3	10.8	1.8	71.6	0.8	14.9	7.90	2.79
0.7	56.6	0.1	0.0	19.7	0.0	0.0	100.0	0.0	0.0	2.96	1.45
1.7	60.2	0.2	0.2	19.7	2.7	0.0	97.3	0.0	0.0	9.00	1.46
2.4	50.6	0.6	0.1	13.7	1.6	0.0	97.9	0.5	0.0	8.51	0.87
3.6	51.8	0.4	0.4	15.2	4.8	0.0	95.2	0.0	0.0	0.24	0.40
0.9	45.2	0.0	0.0	11.8	1.5	0.0	98.5	0.0	0.0	0.25	0.20
1.4	47.4	0.0	0.6	13.3	1.5	0.0	98.5	0.0	0.0	0.24	0.59
3.4	29.5	0.1	0.0	18.5	1.8	12.1	86.0	0.0	0.0	0.24	0.27
4.2	33.4	0.1	1.5	18.0	2.3	11.5	86.2	0.0	0.0	0.24	0.34
4.2	49.2	0.5	3.8	20.4	17.6	2.8	68.6	1.3	9.7	7.51	2.40
2.6	28.7	0.2	0.1	9.9	0.0	5.3	94.7	0.0	0.0	0.25	0.03
6.0	28.4	0.3	6.1	9.0	0.0	8.6	91.4	0.0	0.0	0.25	0.37
8.4	54.4	0.7	16.6	4.9	13.9	0.0	86.1	0.0	0.0	0.23	0.57
2.9	44.3	0.4	0.9	7.0	4.6	6.7	88.7	0.0	0.0	0.24	0.72
4.0	44.1	0.7	5.0	7.8	5.7	8.9	85.4	0.0	0.0	0.24	1.00
4.3	47.0	0.5	6.5	19.0	18.6	3.8	68.0	1.1	8.5	6.60	2.24
4.5	46.5	0.5	9.5	17.9	23.6	3.6	64.0	1.0	7.8	5.98	2.17
3.6	50.5	0.2	18.0	11.0	17.1	26.0	53.0	1.7	2.1	2.40	1.67
3.6	50.4	0.2	18.3	10.9	17.3	25.7	53.2	1.6	2.2	2.37	1.66
10.9	35.4	0.2	0.0	15.6	0.0	0.0	83.9	0.0	16.1	5.90	0.23
26.0	49.7	1.1	0.0	19.6	0.4	3.0	0.0	0.0	96.6	19.37	1.56
26.0	49.7	1.1	0.0	19.6	0.4	3.0	0.0	0.0	96.6	19.37	1.56
12.4	65.7	1.7	0.0	14.7	0.0	6.2	4.8	0.0	89.0	15.15	3.17
3.1	74.3	0.2	0.0	5.2	0.9	0.0	90.9	0.0	8.3	0.12	0.00
22.6	42.9	3.4	8.7	13.5	25.8	8.5	26.6	0.0	39.1	4.19	0.87
2.6	74.2	0.3	0.4	15.3	10.1	3.3	84.4	0.0	2.2	0.25	1.41
3.5	70.3	0.4	5.5	14.8	11.6	2.2	77.5	0.0	8.8	0.26	1.10

Site number	Period of record	Latitude (decimal degrees)	Longitude (decimal degrees)	Drainage area (square miles)	Amount of upstream reservoir storage (acre-feet)	Mean basin elevation (feet)	Mean basin average annual precipitation (inches)	Area covered by developed land (percent)	Area covered by barren land (percent)
09060700	1965–73	40.049708	-106.865878	31.09	0	9,343	31.6	0.0	0.1
09060770	1983–97	39 912209	-106.725594	200.03	0	8,974	26.5	0.4	0.0
09060800	1958–65	39.856098	-106.795873	14.34	0	8,587	18.2	0.0	0.2
09060900	1955–61	39.840542	-106.809484	5.35	0	9,554	25.5	0.0	0.5
09060950	1981–86	39.864431	-106.817541	35.24	0	8,811	20.8	0.0	0.2
09061000	1952–58	39 961931	-106 944491	9.72	0	9,723	35.4	0.0	0.7
09061500	1930–82	39 374155	-106 227801	1.05	0	11,930	28.6	0.0	6.8
09061600	2002 to current year	39.410265	-106 249747	13.43	0	11,480	26.4	0.6	26.3
09062000	1908–82	39 363599	-106 306414	3.11	0	11,362	25.2	0.0	0.3
09063000	1910–25, 1944 to current year	39 508319	-106 366695	75.28	0	10,794	23.4	0.9	5.8
09063200	1964–2008	39.522208	-106 323638	9.62	0	10,832	27.1	0.0	0.6
09063400	1963–2008	39 522764	-106 336139	23.70	0	10,730	28.6	0.0	0.3
09063500	1913–21, 1944–56	39 513875	-106 367250	29.51	0	10,590	27.8	0.0	0.2
09063900	1972 to current year	39 390265	-106.470029	6.48	0	11,380	32.1	0.0	36.2
09064000	1947–54, 1972 to current year	39.405543	-106.433362	36.75	43,600	11,274	30.1	0.0	24.2
09064500	1910–18, 1944 to current year	39.473320	-106 367806	58.27	43,600	10,922	27.7	0.0	16.1
09064600	1989 to current year	39 553875	-106.402529	191.42	43,600	10,725	25.6	0.7	8.5
09065100	1956–63, 1967 to current year	39 568042	-106.412529	34.43	0	11,169	30.7	0.1	30.5
09065500	1947–56, 1963 to current year	39.625819	-106 278082	14.57	0	11,111	30.7	0.2	22.8
09066000	1947–56, 1963 to current year	39 596375	-106 265026	12.52	0	10,701	28.5	4.7	2.3
09066050	1974–79	39.623319	-106 280305	19.52	0	10,521	28.6	4.0	1.5
09066100	1963–2008	39.639986	-106.293360	4.56	0	11,072	31.9	0.0	28.3
09066150	1966–2008	39.643597	-106 302527	5.19	0	11,107	30.9	0.0	24.6
09066200	1964 to current year	39.648319	-106 323083	6.20	0	10,856	28.5	0.7	10.5
09066250	1974–79	39.643042	-106 346139	57.39	0	10,661	29.2	3.2	11.9
09066300	1964 to current year	39.645819	-106 382251	5.96	0	10,466	25.6	0.1	2.9
09066310	1988–99	39.641097	-106 394196	76.92	0	10,514	28.9	3.4	9.2
09066325	1999 to current year	39.641097	-106 394751	76.92	0	10,514	28.9	3.4	9.2
09066400	1963–2008	39.682764	-106.401418	7.32	0	10,394	26.7	0.0	3.5
09066500	1911–14, 1944–56	39.614708	-106.440030	100.98	0	10,301	27.9	3.6	7.2
09066510	1995 to current year	39.609430	-106.447808	101.87	0	10,287	27.8	3.6	7.2
09067000	1911, 1912–14, 1974–87, 1988 to current year	39.629708	-106 522810	14.81	0	10,196	27.1	4.1	4.5
09067005	1988–99	39.631653	-106 522532	399.01	43,600	10,479	26.8	1.9	8.8
09067020	1999 to current year	39.635000	-106 532500	406.14	43,600	10,455	26.7	2.0	8.7
09067200	1993 to current year	39.647486	-106.609200	46.63	0	10,311	27.9	0.1	19.5
09067300	1958–65	39.756098	-106.670591	27.12	0	8,548	18.3	0.6	0.1
09067500	1910–24	39.656652	-106.825317	634.75	43,600	9,853	24.4	2.5	7.0
09067700	1965–72	39 504152	-106.678091	9.75	0	11,029	31.1	0.0	24.8
09068000	1950–72	39 557207	-106.763093	71.52	0	9,952	26.1	0.0	6.2
09069500	1950–55, 1965–72	39 545540	-106 934766	62.77	0	9,695	24.6	0.0	3.2
09070000	1946 to current year	39.649430	-106 953655	950.25	43,600	9,494	23.1	2.0	5.4
09070500	1940 to current year	39.643873	-107.077826	4,392.05	1,118,665	9,386	24.0	1.2	4.1
09071100	1941–85	39 569983	-107 226720	4,485.17	1,118,665	9,363	24.0	1.2	4.0
09071300	1976–96	39.716649	-107 310333	6.73	0	10,774	42.7	0.0	0.0
09072500	1899–1966	39 549982	-107 320887	4,559.40	1,118,665	9,364	24.1	1.2	3.9
09072550	1980 to current year	39 120268	-106.624755	9.19	0	11,945	28.9	3.3	20.5
09073005	1980 to current year	39.079991	-106.616421	15.15	0	12,005	31.4	0.0	38.9
09073300	1979 to current year	39 141100	-106.774204	75.86	0	11,525	29.2	1.3	22.5
09073400	1964 to current year	39 179988	-106.801983	106.79	0	11,250	28.8	1.2	16.5
09073500	1910–21, 1931–64	39.189432	-106.814483	108.64	0	11,211	28.7	1.5	16.2
09073700	1964–80	39.213877	-106.655867	6.14	0	11,754	32.0	0.0	39.9
09073720	1981–83	39 207766	-106.678645	8.42	0	11,578	31.4	0.0	32.1

Area covered by deciduous forest (percent)	Area covered by evergreen forest (percent)	Area covered by mixed forest (percent)	Area covered by shrubs young or stunted trees (percent)	Area covered by grass or herbaceous land (percent)	Sedimentary, clastic lithology, Mesozoic (percent)	Sedimentary, clastic (continental) lithology, Tertiary (percent)	Igneous or metamorphic lithologies (percent)	Sedimentary, mixed (continental and marine) lithology (percent)	Sedimentary, carbonate (marine) lithology (percent)	Population density (people per square mile)	Road density (miles per square mile)
26.7	26.8	7.2	12.8	22.7	59.5	12.3	28.2	0.0	0.0	0.32	0.96
13.6	43.2	2.3	20.2	13.9	43.7	6.4	39.9	0.0	9.9	0.29	1.08
19.5	17.3	3.0	53.3	6.2	62.3	14.6	23.2	0.0	0.0	0.12	1.60
23.1	47.2	5.3	19.6	4.3	55.5	36.4	8.1	0.0	0.0	0.19	0.65
20.2	21.4	2.7	43.5	11.0	65.2	23.4	11.4	0.0	0.0	0.21	1.29
26.4	26.6	3.7	4.1	37.2	75.4	7.0	17.6	0.0	0.0	0.34	0.36
0.0	14.5	0.0	0.0	71.2	0.0	0.0	0.0	0.0	99.2	0.89	0.55
0.7	23.5	0.0	0.0	41.2	0.0	0.5	25.6	0.0	73.8	0.77	2.04
0.2	60.9	0.0	0.0	37.2	0.0	0.0	60.8	0.0	39.2	0.89	0.53
7.2	57.8	0.2	0.2	23.3	0.0	0.1	36.6	3.3	59.9	0.87	1.63
6.9	72.1	0.1	0.1	19.4	0.0	8.6	6.1	0.0	85.3	0.88	2.30
8.9	71.8	0.4	0.2	17.1	0.0	5.6	6.4	0.0	88.0	0.88	2.20
11.2	70.6	0.7	0.5	15.5	0.0	4.5	5.2	0.0	90.4	0.89	2.13
0.0	37.8	0.0	0.0	18.5	0.0	0.0	100.0	0.0	0.0	0.85	0.37
1.5	43.8	0.1	0.0	23.8	0.0	1.2	98.8	0.0	0.0	0.70	0.60
6.8	51.4	0.5	0.4	19.5	0.0	0.7	98.5	0.0	0.7	0.77	0.62
8.4	58.2	0.4	0.6	19.0	0.0	1.0	56.4	1.3	41.4	0.84	1.35
4.7	36.8	0.5	1.0	19.8	0.0	0.0	96.3	0.0	3.7	0.89	0.04
5.4	36.3	0.1	0.3	24.2	0.0	0.0	98.4	0.0	1.6	3.24	0.09
4.9	54.1	0.0	0.0	30.5	0.0	6.2	28.0	0.0	65.7	2.56	1.24
7.2	62.2	0.6	0.0	22.0	0.0	15.0	21.9	0.0	63.1	2.38	1.29
8.0	27.1	0.3	0.4	28.6	0.0	0.0	99.7	0.0	0.3	3.23	0.00
6.6	28.0	0.4	0.0	31.0	0.0	0.0	92.2	0.0	7.8	3.22	0.00
9.9	33.6	0.4	1.5	34.4	0.0	0.0	65.7	1.6	32.6	3.27	0.08
9.9	43.2	1.6	1.2	22.8	0.0	5.3	58.0	1.1	35.5	2.95	0.88
11.2	70.8	3.5	2.1	7.7	0.0	0.0	29.7	0.3	70.0	5.85	0.55
11.9	46.3	2.0	1.9	20.5	0.0	6.8	46.7	1.9	44.5	4.15	1.42
11.9	46.3	2.0	1.9	20.5	0.0	6.8	46.7	1.9	44.5	4.15	1.42
3.0	77.4	0.0	0.0	13.3	0.0	0.0	55.0	0.0	45.0	7.88	5.50
14.1	48.0	2.4	2.7	17.6	0.0	6.2	40.2	2.0	51.6	5.68	1.97
14.5	47.7	2.4	2.9	17.4	0.0	6.2	39.9	2.0	52.0	5.72	1.99
17.6	50.7	1.8	5.4	14.8	0.0	0.0	58.2	0.0	41.8	13.52	2.01
11.7	51.5	1.5	2.7	17.8	0.0	3.1	49.5	1.5	45.8	4.55	1.46
12.0	51.0	1.6	3.2	17.6	0.0	3.2	48.6	1.6	46.6	4.99	1.48
17.4	41.2	2.4	6.6	9.2	0.0	0.0	71.5	3.6	24.9	6.19	0.85
14.4	1.6	0.4	67.9	14.3	58.3	32.7	8.9	0.0	0.0	0.12	2.24
16.8	39.7	1.8	14.9	13.9	12.2	5.1	38.0	2.0	42.7	5.31	1.82
9.9	46.6	0.3	0.0	16.1	0.0	0.0	81.4	0.0	18.6	4.21	0.68
26.1	48.2	2.4	4.3	11.6	1.5	0.0	27.8	0.0	70.7	4.06	1.68
27.1	41.4	3.4	7.4	16.6	0.0	0.0	22.5	0.0	77.5	4.24	0.80
18.9	37.6	1.9	18.5	12.2	11.9	3.9	30.4	3.2	50.5	4.80	1.87
9.4	44.3	0.9	20.8	11.9	19.5	17.4	42.4	1.6	19.1	2.50	1.61
10.0	43.7	1.0	20.9	11.9	19.1	17.2	41.7	1.6	20.3	2.48	1.60
0.5	11.9	0.0	0.0	79.2	0.0	0.0	4.6	0.0	95.4	0.62	1.48
10.2	43.5	1.0	20.6	12.2	18.8	16.9	41.3	1.5	21.4	2.47	1.60
0.1	17.4	0.0	0.0	49.9	0.0	0.0	99.9	0.0	0.0	0.16	0.62
0.1	16.5	0.0	0.0	37.6	0.0	0.0	99.9	0.0	0.0	0.16	0.34
3.6	30.9	0.3	0.0	36.0	0.0	0.0	100.0	0.0	0.0	0.16	0.38
6.0	38.6	0.4	0.1	31.9	0.0	0.0	97.6	0.0	2.4	1.04	0.45
6.5	38.6	0.5	0.1	31.4	0.0	0.0	97.2	0.1	2.6	2.40	0.57
0.2	26.5	0.0	0.0	28.0	0.0	0.0	100.0	0.0	0.0	0.16	0.00
0.8	38.3	0.0	0.0	24.1	0.0	0.0	100.0	0.0	0.0	0.16	0.00

Site number	Period of record	Latitude (decimal degrees)	Longitude (decimal degrees)	Drainage area (square miles)	Amount of upstream reservoir storage (acre-feet)	Mean basin elevation (feet)	Mean basin average annual precipitation (inches)	Area covered by developed land (percent)	Area covered by barren land (percent)
09073790	1981–83	39.190266	-106.685034	8.04	0	11,491	30.2	0.0	22.4
09073800	1971–80	39 196377	-106.690035	8.68	0	11,424	30.0	0.0	20.8
09073890	1981–83	39 182766	-106.720035	5.68	0	11,308	29.1	0.0	3.7
09073900	1971–80	39 188877	-106.718369	6.72	0	11,251	29.1	0.0	3.7
09074000	1950–56, 1969 to current year	39 205820	-106.797538	40.80	0	10,885	29.1	0.0	11.8
09074800	1969–94	39.087489	-106.812261	31.98	0	11,434	36.4	0.0	27.3
09075000	1911–20	39 139710	-106.846984	65.16	0	11,224	36.6	0.0	25.6
09075500	1913–18	39 210820	-106.839762	226.24	0	11,051	30.9	1.4	17.3
09075700	1969–94	39 123599	-106 905320	35.56	0	11,373	43.7	0.0	34.0
09076000	1910–17	39 163043	-106.880874	41.87	0	11,171	41.5	0.0	29.9
09076520	1974–89	39.223597	-106.879763	6.48	0	9,014	24.7	0.0	0.0
09077150	1971–83	39 244988	-106 530308	10.37	0	11,937	37.3	0.0	43.4
09077200	1963–82	39 245544	-106 531419	18.85	0	11,847	36.2	0.0	39.2
09077400	1963–67	39 296098	-106.603643	32.18	0	11,381	34.1	0.1	26.3
09077600	1963–76	39 289154	-106 553086	9.11	0	11,367	34.0	0.0	8.7
09077605	1976–83	39 287210	-106 559198	9.26	0	11,353	34.0	0.0	8.8
09077610	1976–82	39 286932	-106 559198	9.46	0	11,345	33.9	0.0	9.5
09077800	1963–82	39 241654	-106 593087	11.84	0	11,593	33.2	0.0	31.0
09077900	1963–67	39 301931	-106.619199	17.34	0	11,180	31.9	0.0	23.4
09077940	1971–83	39 246376	-106.629755	4.69	0	11,611	33.3	0.0	32.2
09077945	1973–82	39 264154	-106.632255	6.24	0	11,436	32.9	0.0	26.6
09077950	1966–72	39 267209	-106.635033	6.35	0	11,413	32.9	0.0	26.1
09078000	1910–17, 1947–83	39 330820	-106.658090	89.06	0	10,973	31.7	0.1	17.8
09078040	1980–83	39.361654	-106 537809	1.50	0	11,710	33.7	0.0	40.4
09078050	1979–83	39 355543	-106 533920	3.20	0	11,286	33.7	0.0	24.9
09078060	1980–83	39 366654	-106 544476	2.91	0	11,571	34.4	0.0	48.5
09078100	1963–80	39 358876	-106 568365	11.17	0	11,160	32.7	0.0	26.4
09078140	1979–83	39 336654	-106 542531	1.85	0	11,206	33.7	0.0	21.6
09078150	1980–83	39 328598	-106 552253	1.99	0	11,111	33.6	0.0	10.4
09078200	1963–80	39 334154	-106 575310	7.27	0	10,875	32.0	0.0	11.7
09078300	1963–68	39 343598	-106.611977	24.61	0	10,777	31.2	0.0	15.9
09078500	1910–17, 1947–82	39 342764	-106.665868	41.69	0	10,538	29.7	0.0	12.0
09078600	1975–95	39 344708	-106.673646	132.98	0	10,805	30.9	0.0	15.7
09078900	1963–68	39.432764	-106 596144	4.56	0	11,597	34.1	0.0	58.7
09079000	1950–56	39.422486	-106.639200	7.79	0	11,117	32.1	0.0	39.6
09079500	1950–56	39 354430	-106.693368	35.15	0	10,216	28.0	0.0	9.8
09080000	1915–20	39.363041	-106.705035	172.88	0	10,634	30.0	0.0	14.1
09080100	1910–15, 1966–80	39.362486	-106.732536	189.49	0	10,565	29.7	0.0	12.9
09080200	1959–64	39 361374	-106.819206	223.86	102,500	10,402	28.7	0.0	11.0
09080300	1968–82	39 361652	-106.820595	236.24	102,500	10,386	28.6	0.0	10.5
09080400	1964 to current year	39 365541	-106.825595	236.51	102,500	10,384	28.6	0.0	10.5
09080800	1963–68	39 336373	-107.095603	14.84	0	9,000	27.8	0.0	1.8
09081000	1908–09, 1998 to current year	39 373317	-107.083937	858.62	102,500	10,118	29.0	1.2	10.9
09081500	1910–15, 1916–17	39.068600	-107 190604	73.80	0	11,018	45.0	0.6	18.2
09081550	1959–73, 1975–77	39 142764	-107 257828	107.32	0	10,566	42.0	0.8	16.9
09081600	1955 to current year	39 232207	-107 227273	167.42	0	10,164	39.5	0.7	12.3
09082500	1935–63	39 298595	-107 214217	229.50	0	10,083	38.6	0.6	13.6
09082800	1963–79	39 329705	-107 333386	27.91	0	9,454	33.9	0.0	0.0
09083000	1950–60, 1964–68	39 330539	-107.224496	75.10	0	9,167	32.4	0.0	0.0
09083700	1963–68	39 323873	-107.139493	2.79	0	9,797	35.8	0.0	27.9
09083800	2000 to current year	39.408039	-107 230330	349.13	0	9,556	34.9	0.8	9.3
09084000	1950–55, 1962–72	39.466650	-107.052270	30.49	0	9,493	27.1	0.0	0.6
09084500	1941–47	39 395537	-107 388942	7.74	0	9,354	32.7	0.0	0.0
09084600	1957–65	39.400538	-107 324497	16.56	0	9,225	30.8	0.2	0.0

Area covered by deciduous forest (percent)	Area covered by evergreen forest (percent)	Area covered by mixed forest (percent)	Area covered by shrubs young or stunted trees (percent)	Area covered by grass or herbaceous land (percent)	Sedimentary, clastic lithology, Mesozoic (percent)	Sedimentary, clastic (continental) lithology, Tertiary (percent)	Igneous or metamorphic lithologies (percent)	Sedimentary, mixed (continental and marine) lithology (percent)	Sedimentary, carbonate (marine) lithology (percent)	Population density (people per square mile)	Road density (miles per square mile)
0.4	44.1	0.0	0.0	30.0	0.0	0.0	100.0	0.0	0.0	0.15	0.00
0.3	47.3	0.1	0.0	28.5	0.0	0.0	100.0	0.0	0.0	0.15	0.00
0.0	59.1	0.0	0.0	31.2	0.0	0.0	100.0	0.0	0.0	0.15	0.00
0.0	63.1	0.0	0.0	28.1	0.0	0.0	100.0	0.0	0.0	0.15	0.00
10.0	54.2	0.6	0.3	19.8	0.0	0.0	95.1	0.0	4.9	0.88	0.32
12.8	28.4	0.4	0.0	28.5	0.0	0.0	55.6	0.0	44.4	0.15	0.86
15.5	30.6	1.0	0.1	25.0	0.0	0.0	45.4	0.0	54.6	1.53	0.71
11.4	38.4	0.8	0.3	26.3	0.2	0.0	78.3	0.8	20.7	3.68	0.84
11.9	21.2	0.6	0.0	28.7	0.0	0.0	22.2	0.0	77.8	3.88	0.12
15.7	23.9	0.8	0.0	26.6	0.3	0.0	18.8	0.0	80.8	4.01	0.17
52.1	27.5	1.1	12.4	2.3	67.4	0.0	28.2	0.0	4.4	2.38	1.31
0.6	19.9	0.0	0.0	30.0	0.0	0.0	99.7	0.0	0.0	0.35	0.00
0.4	20.4	0.0	0.0	34.1	0.0	0.0	99.8	0.0	0.0	0.35	0.01
1.7	40.3	0.2	0.0	27.1	0.0	0.0	99.9	0.0	0.0	0.35	0.27
0.1	34.2	0.0	0.0	51.2	0.0	0.0	99.9	0.0	0.0	0.37	1.13
0.1	34.2	0.0	0.0	51.2	0.0	0.0	99.9	0.0	0.0	0.37	1.16
0.1	34.6	0.0	0.0	50.2	0.0	0.0	99.9	0.0	0.0	0.37	1.13
0.5	37.8	0.0	0.0	25.5	0.0	0.0	100.0	0.0	0.0	0.33	0.01
1.3	48.7	0.6	0.0	22.1	0.0	0.0	100.0	0.0	0.0	0.34	0.08
0.0	33.7	0.0	0.0	30.2	0.0	0.0	100.0	0.0	0.0	0.34	0.00
0.5	40.6	0.0	0.0	29.3	0.0	0.0	100.0	0.0	0.0	0.34	0.07
0.5	41.6	0.0	0.0	28.8	0.0	0.0	100.0	0.0	0.0	0.34	0.14
5.6	47.6	0.7	0.9	23.1	0.0	0.0	99.1	0.0	0.9	0.35	0.55
0.0	23.5	0.0	0.0	30.9	0.0	0.0	100.0	0.0	0.0	0.37	0.00
0.3	49.7	0.0	0.0	20.6	0.0	0.0	100.0	0.0	0.0	0.35	0.00
0.0	17.3	0.0	0.0	26.7	0.0	0.0	100.0	0.0	0.0	0.45	0.00
1.4	47.8	0.0	0.0	19.8	0.0	0.0	100.0	0.0	0.0	0.38	0.15
0.4	49.9	0.0	0.0	24.8	0.0	0.0	100.0	0.0	0.0	0.35	0.00
0.1	69.5	0.0	0.0	19.4	0.0	0.0	100.0	0.0	0.0	0.35	0.40
1.9	68.6	0.0	0.0	16.5	0.0	0.0	100.0	0.0	0.0	0.35	0.47
5.3	60.1	0.3	0.2	15.6	0.0	0.0	100.0	0.0	0.0	0.37	0.46
10.7	58.9	0.7	1.5	13.5	0.0	0.0	98.6	0.0	1.4	0.37	0.55
7.5	51.2	0.8	1.3	19.7	0.0	0.0	98.9	0.0	1.1	0.35	0.56
0.3	14.2	0.0	0.0	21.0	0.0	0.0	100.0	0.0	0.0	0.66	0.65
8.0	27.4	0.1	0.0	18.6	0.0	0.0	89.2	0.0	10.8	0.61	0.81
23.2	46.5	2.2	1.1	14.1	0.0	0.0	42.4	0.0	57.6	0.45	1.38
11.7	49.6	1.2	1.7	18.1	0.0	0.3	85.5	0.0	14.2	0.37	0.75
12.6	50.0	1.3	2.3	17.4	0.0	1.0	80.5	0.0	18.5	0.37	0.73
15.8	47.6	1.5	3.3	17.0	0.0	2.3	69.4	0.0	28.3	0.38	0.77
15.5	49.0	1.5	3.2	16.7	0.0	2.2	67.1	0.0	30.6	0.37	0.84
15.5	49.0	1.5	3.2	16.7	0.0	2.2	67.1	0.0	30.7	0.37	0.84
52.0	20.6	4.1	12.4	6.7	67.4	0.6	23.6	0.0	8.4	5.24	1.48
21.6	36.1	1.5	7.8	16.8	12.4	1.9	47.3	2.6	35.8	2.91	1.06
14.2	24.5	1.0	0.3	33.7	44.2	0.5	19.6	0.0	35.7	0.13	0.59
22.4	24.5	2.3	1.3	26.0	40.9	6.1	28.5	0.0	24.5	0.14	0.73
27.9	28.4	2.9	2.2	21.3	46.9	4.0	20.0	0.0	29.2	0.29	0.76
26.1	30.1	3.0	3.0	19.9	34.9	2.9	27.1	0.0	35.1	0.37	0.63
41.9	41.2	6.9	0.7	8.8	3.8	96.2	0.0	0.0	0.0	0.65	1.25
38.4	42.8	4.8	5.6	7.9	43.4	51.7	0.0	0.0	4.9	0.61	0.85
25.5	25.5	8.1	5.3	6.2	25.5	0.0	63.3	0.0	11.2	2.54	0.83
29.8	31.6	3.2	6.4	14.9	35.6	17.3	19.2	0.9	27.0	1.52	0.86
41.8	36.9	7.9	3.2	9.2	34.3	0.0	23.8	0.0	41.9	0.74	1.53
53.1	24.5	10.6	0.2	9.7	0.0	100.0	0.0	0.0	0.0	2.50	1.77
67.5	13.8	7.2	1.7	8.3	3.2	96.8	0.0	0.0	0.0	2.64	1.57

Site number	Period of record	Latitude (decimal degrees)	Longitude (decimal degrees)	Drainage area (square miles)	Amount of upstream reservoir storage (acre-feet)	Mean basin elevation (feet)	Mean basin average annual precipitation (inches)	Area covered by developed land (percent)	Area covered by barren land (percent)
09085000	1905–09, 1910 to current year	39.543593	-107 329498	1,452.53	102,500	9,598	29.3	1.3	8.7
09085100	1966 to current year	39 554982	-107 337554	6,014.60	1,221,165	9,419	25.4	1.2	5.1
09085200	1969–86	39.605259	-107.448388	4.89	0	9,474	31.3	0.0	0.0
09085300	1969–83	39.609148	-107.434777	15.50	0	9,626	33.9	0.0	0.5
09085400	1969–82	39 597759	-107.423943	6.55	0	8,956	28.7	0.0	0.0
09085500	1954–60	39 574426	-107.447554	55.44	0	9,224	31.9	0.1	0.4
09086000	1991–97	39.666369	-107.627004	9.51	0	8,180	24.1	0.0	0.0
09086470	1991–97	39.678036	-107 573114	90.35	0	9,319	32.8	0.2	0.2
09086970	1991–97	39.668036	-107 525056	23.83	0	10,090	37.9	0.0	0.2
09087500	1922–24, 1954–60	39 570536	-107 539778	178.46	0	8,641	28.9	0.5	0.2
09087600	1966–72	39 568314	-107 541167	6,306.36	1,221,165	9,375	25.5	1.2	4.9
09088000	1955–61	39.486092	-107 501442	15.42	0	8,778	28.4	0.0	0.0
09089000	1938–47, 1963–70	39 275537	-107 520055	35.08	0	9,046	32.2	0.0	0.0
09089500	1955–2005	39 331092	-107 580056	64.27	0	8,732	30.1	0.0	0.0
09090700	1959–65	39.436370	-107 569500	40.59	0	8,595	28.5	0.0	0.0
09091500	1936–43, 1956–64	39.677758	-107.698396	35.00	0	8,552	26.1	1.1	0.0
09092000	1939–46, 1952–64	39.619980	-107.763398	137.09	13,600	7,970	22.9	0.5	0.1
09092500	1952–82	39.471924	-107.832566	11.96	0	8,881	26.6	0.0	1.5
09092600	1956–65	39.435534	-107 977294	9.82	0	9,156	29.0	0.0	4.9
09092800	1957–62	39 586921	-108 143413	5.35	0	8,108	19.9	0.0	2.9
09092830	1976–83	39.620255	-108.012852	12.57	0	8,606	24.4	0.0	0.3
09092850	1976–83	39.620810	-108.030075	22.14	0	8,521	24.0	0.0	0.6
09092960	1976–83	39 554978	-107.982851	14.43	0	8,753	24.8	0.0	0.1
09092970	1976–83	39 567477	-108.021186	20.33	0	8,616	24.3	0.0	0.3
09092980	1976–83	39 590255	-108.041187	4.05	0	8,287	23.1	0.0	1.7
09093000	1948–54, 1964–70, 1975–86	39 566922	-108 110912	140.50	0	8,129	21.2	0.1	3.3
09093500	1921–27, 1948–54, 1975–82	39.453034	-108.059798	196.71	0	7,849	20.5	0.2	5.3
09093700	1967–97	39 362479	-108 152579	7,363.42	1,239,823	9,091	24.8	1.3	4.5
09094200	1962–68	39.491088	-108 355087	151.85	0	7,678	20.9	0.1	7.1
09094400	1966–68	39.483311	-108 326753	111.02	0	7,849	20.4	0.2	5.0
09095000	1921–26, 1962–72, 1975–81	39.453311	-108 317030	322.30	0	7,623	20.3	0.2	6.5
09095300	1996–98, 2001–2004	39 374701	-108 317864	97.07	0	7,026	17.5	0.0	6.3
09095400	1974–82	39 368868	-108 262028	108.45	0	6,893	17.1	0.0	5.8
09095500	1933 to current year	39 239146	-108.266195	7,986.47	1,239,823	8,929	24.3	1.2	4.6
09095800	1958–64	39 217480	-107.774506	18.41	0	9,639	32.2	0.0	0.0
09096500	1921–80	39 250535	-107.840620	80.46	33,800	9,744	34.3	0.0	1.2
09096800	1955–70	39.236092	-107.633946	50.21	0	9,338	30.1	0.5	0.0
09097500	1921–80	39 272202	-107.850621	9.33	0	9,588	31.0	0.2	0.2
09097600	1955–67	39 324979	-107.842287	9.33	0	9,588	31.0	0.2	0.2
09097900	2003 to current year	39 239701	-107 971459	327.51	33,800	8,698	27.3	0.7	0.4
09098500	1952–55	39 102758	-107.897288	1.00	0	10,141	37.5	0.0	0.0
09099000	1940–41, 1950–55	39 109424	-107.895621	4.89	0	10,384	38.9	0.0	2.6
09099500	1945–56	39 131924	-107 918678	19.90	0	10,137	37.7	0.0	1.7
09100000	1937–44	39 194979	-107 961180	26.79	0	9,804	35.3	0.1	1.3
09100500	1945–57	39 127757	-107 996459	14.39	0	9,903	35.4	0.0	3.5
09101000	1937–43	39 164146	-108.018405	18.73	0	9,492	33.0	0.0	2.7
09101500	1945–53	39 116924	-108.035627	10.54	0	9,936	34.8	0.0	1.6
09104000	1937–43	39 106647	-108.133410	8.09	0	9,341	30.9	1.2	0.7
09104500	1937–60	39.086369	-108.126742	5.47	0	9,836	33.6	2.5	3.3
09105000	1935–83, 1985 to current year	39 183314	-108.267861	593.40	40,120	8,203	25.3	0.8	0.9
09106000	1901–33	39 127759	-108 325918	8,735.61	1,279,943	8,833	24.2	1.2	4.3
09106150	1990 to current year	39.098592	-108 355086	8,751.16	1,279,943	8,828	24.2	1.2	4.3
09106200	1973–79, 2002–04	39.060537	-108.477868	5.49	0	5,087	10.3	10.9	3.2
09107000	1929–34, 1987 to current year	38.860271	-106 566697	127.90	0	10,926	26.4	0.0	11.1

Area covered by deciduous forest (percent)	Area covered by evergreen forest (percent)	Area covered by mixed forest (percent)	Area covered by shrubs young or stunted trees (percent)	Area covered by grass or herbaceous land (percent)	Sedimentary, clastic lithology, Mesozoic (percent)	Sedimentary, clastic (continental) lithology, Tertiary (percent)	Igneous or metamorphic lithologies (percent)	Sedimentary, mixed (continental and marine) lithology (percent)	Sedimentary, carbonate (marine) lithology (percent)	Population density (people per square mile)	Road density (miles per square mile)
26.2	32.1	2.0	10.8	14.0	18.6	8.3	38.4	2.7	32.0	3.72	1.31
14.1	40.7	1.3	18.3	12.7	18.8	14.8	40.6	1.8	24.0	2.81	1.53
42.8	30.5	2.2	3.4	20.6	0.0	0.0	5.9	0.0	94.1	0.62	1.81
25.1	36.2	3.5	5.0	26.6	0.0	0.0	23.2	0.0	76.8	0.62	0.69
50.3	24.4	1.9	7.3	15.3	0.0	0.0	4.4	0.0	95.6	0.62	1.43
28.4	30.1	2.1	9.3	25.9	0.0	2.5	17.3	0.0	80.2	1.09	0.89
75.0	13.5	2.0	2.5	4.3	11.2	0.0	0.0	0.0	88.8	1.67	1.24
33.2	26.2	4.8	4.8	29.3	0.0	0.0	1.2	0.0	98.8	2.10	0.78
14.1	45.7	1.3	1.0	36.7	0.0	0.0	6.6	0.0	93.4	2.16	0.55
28.1	32.7	2.9	10.4	20.6	14.7	0.1	1.5	0.0	83.8	2.06	1.06
14.9	40.3	1.3	18.0	12.9	18.6	14.3	38.9	1.7	26.4	2.79	1.52
76.2	9.4	7.3	4.0	2.8	0.0	100.0	0.0	0.0	0.0	0.94	1.22
73.4	15.8	3.3	2.8	4.3	5.8	93.8	0.4	0.0	0.0	0.39	0.93
75.7	9.9	2.3	7.7	3.6	7.6	92.2	0.2	0.0	0.0	0.39	0.89
67.1	15.6	6.7	7.4	2.3	0.5	99.5	0.0	0.0	0.0	0.74	1.32
56.3	18.7	7.0	3.5	12.6	0.0	4.6	0.0	0.0	95.4	0.47	1.18
40.3	29.6	3.2	14.6	9.6	40.5	7.4	0.1	0.0	51.9	0.80	1.13
50.8	27.1	1.7	9.5	8.9	10.9	84.7	4.4	0.0	0.0	0.46	1.57
33.0	42.3	11.0	4.3	3.2	0.0	90.7	9.3	0.0	0.0	8.04	0.51
33.1	3.3	0.0	60.0	0.5	0.0	100.0	0.0	0.0	0.0	0.15	1.58
48.2	15.4	0.5	29.0	6.6	0.0	100.0	0.0	0.0	0.0	0.16	1.42
41.5	16.4	0.3	35.5	5.6	0.0	100.0	0.0	0.0	0.0	0.16	1.48
57.0	5.9	0.2	30.0	6.7	0.0	100.0	0.0	0.0	0.0	0.17	0.89
57.2	6.3	0.2	30.5	5.3	0.0	100.0	0.0	0.0	0.0	0.16	0.78
50.9	14.7	0.0	29.0	3.2	0.0	100.0	0.0	0.0	0.0	0.15	0.86
37.6	13.6	0.1	41.9	2.3	0.0	100.0	0.0	0.0	0.0	0.16	1.45
33.1	16.3	0.1	40.5	1.8	0.0	98.7	0.0	1.3	0.0	0.16	1.51
17.4	38.1	1.3	19.3	11.4	17.4	23.4	33.4	2.2	23.5	2.79	1.55
19.9	35.9	0.1	30.8	2.5	0.0	100.0	0.0	0.0	0.0	0.15	1.02
23.4	13.6	0.0	54.8	0.4	0.0	100.0	0.0	0.0	0.0	0.15	1.26
20.3	29.5	0.1	38.5	1.5	0.0	100.0	0.0	0.0	0.0	0.15	1.03
16.1	44.8	0.0	28.4	1.3	0.0	100.0	0.0	0.0	0.0	0.25	0.84
14.4	45.4	0.0	30.3	1.2	0.0	100.0	0.0	0.0	0.0	0.25	0.82
17.2	37.6	1.2	21.0	10.6	16.4	29.0	30.8	2.1	21.7	2.59	1.51
39.7	42.7	4.2	0.5	10.4	7.7	44.4	47.9	0.0	0.0	0.39	0.12
30.3	44.0	1.9	1.9	15.6	4.4	28.6	67.0	0.0	0.0	0.39	0.49
56.6	25.9	4.1	4.2	7.5	0.0	75.6	23.2	1.2	0.0	0.39	0.63
43.8	26.2	3.6	1.7	24.2	0.0	100.0	0.0	0.0	0.0	0.40	0.27
43.8	26.2	3.6	1.7	24.2	0.0	100.0	0.0	0.0	0.0	0.40	0.27
48.2	20.9	2.3	13.0	7.5	1.7	68.6	27.7	1.9	0.0	0.39	0.81
4.0	73.7	1.2	0.0	5.8	0.0	0.0	100.0	0.0	0.0	0.39	0.71
3.3	61.1	0.5	0.0	26.3	0.0	0.0	100.0	0.0	0.0	0.39	0.36
15.8	57.4	3.9	0.0	15.6	0.0	5.5	94.5	0.0	0.0	0.42	0.82
23.7	49.7	6.8	0.7	12.9	0.0	7.2	92.8	0.0	0.0	0.42	0.95
24.3	51.6	11.1	0.3	4.6	0.0	13.0	87.0	0.0	0.0	0.43	0.89
31.3	43.8	11.5	1.8	3.8	0.0	20.8	79.2	0.0	0.0	0.43	0.94
18.6	58.5	14.3	0.0	3.3	0.0	6.7	93.3	0.0	0.0	0.42	0.54
46.9	22.8	9.2	6.2	5.9	0.0	13.7	86.3	0.0	0.0	0.42	1.46
28.7	49.2	6.3	0.4	3.7	0.0	6.1	93.9	0.0	0.0	0.42	1.54
35.6	28.7	3.0	16.7	5.5	4.2	61.3	31.2	3.3	0.0	0.41	1.01
18.2	37.2	1.3	21.1	10.1	16.6	31.1	30.3	2.1	19.8	2.40	1.47
18.2	37.2	1.3	21.1	10.1	16.7	31.1	30.3	2.1	19.8	2.43	1.47
0.0	0.6	0.0	79.2	0.0	86.0	0.1	0.0	13.9	0.0	58.51	2.56
1.5	53.0	0.0	0.2	27.6	0.0	0.0	91.9	2.8	5.3	0.12	0.70

Site number	Period of record	Latitude (decimal degrees)	Longitude (decimal degrees)	Drainage area (square miles)	Amount of upstream reservoir storage (acre-feet)	Mean basin elevation (feet)	Mean basin average annual precipitation (inches)	Area covered by developed land (percent)	Area covered by barren land (percent)
09108000	1913–14, 1929–34	38.816105	-106 529473	59.20	0	10,968	22.3	0.5	7.9
09109000	1929–34, 1938 to current year	38.818327	-106.609198	254.46	106,200	10,875	24.5	0.3	11.4
09110000	1910 to current year	38.664437	-106.845317	477.40	106,200	10,647	23.9	0.3	8.1
09110500	1939–51	38.864437	-106 909764	89.38	0	10,880	38.3	0.4	15.6
09111000	1941–46	38.856382	-107.055881	8.79	0	10,563	32.9	1.2	0.9
09111500	1940–51, 1993–2006	38.869715	-106 969489	68.33	0	10,333	33.2	1.5	7.8
09112000	1910–13, 1940–51	38.824437	-106.852817	33.04	0	10,807	32.0	0.0	5.8
09112200	1963–72, 1979–81, 1993 to current year	38.784160	-106.870874	240.93	0	10,473	33.6	1.0	9.5
09112500	1905, 1910–22, 1934 to current year	38.664437	-106.848095	288.85	0	10,271	31.6	1.0	8.0
09113000	1944–50	38.766383	-107 101438	20.06	0	10,597	33.8	0.0	12.7
09113100	1993–98	38.769161	-107.084493	22.00	0	10,503	32.9	0.0	11.5
09113300	1958–70	38.765550	-107.058301	47.41	0	10,194	31.4	0.0	8.1
09113500	1940–50, 1958–71, 1979–81	38.702216	-106 998379	119.36	0	9,892	27.0	0.1	5.1
09113980	1998 to current year	38 587770	-106 931432	160.56	0	9,566	24.2	0.4	3.8
09114000	1944–50	38 575269	-106 938655	163.20	0	9,540	23.9	0.4	3.8
09114500	1910–28, 1944 to current year	38 541936	-106 949766	1,010.15	106,200	10,210	25.5	0.6	6.7
09115500	1916–22, 1937–72, 1992 to current year	38.411663	-106.422806	148.45	0	10,234	21.8	0.6	1.9
09116000	1944–50	38.412495	-106 507253	194.89	0	10,088	21.3	0.7	1.6
09117000	1944–51, 1963–70	38.497214	-106.726147	426.73	0	9,605	18.5	0.7	0.9
09118000	1937–50, 1959–70	38 559715	-106.636422	106.94	0	10,617	23.5	0.1	6.3
09118450	1981 to current year	38 335549	-106.772260	333.76	0	10,198	19.8	0.3	1.9
09118500	1940–48	38 399992	-106.765315	360.08	0	10,116	19.3	0.3	1.8
09119000	1910, 1937 to current year	38 521658	-106 940877	1,058.59	0	9,735	18.7	0.6	1.6
09121500	1946–54	37 981108	-107 168663	23.26	0	11,410	28.2	0.7	3.3
09121800	1960–63	38 227495	-107.073383	246.25	0	10,686	22.3	0.5	4.1
09122000	1937–55	38 291383	-107 114496	340.12	0	10,433	20.9	0.5	3.5
09122500	1955–66	38 560825	-107 316722	58.39	0	9,862	27.1	0.0	2.7
09123400	1981–86	37 906386	-107 384779	57.37	0	12,047	41.5	0.0	32.5
09123500	1917–24, 1928–30, 1931–37	38.024997	-107 308389	123.48	0	11,418	33.7	0.4	20.7
09124000	1917–19, 1928–30, 1931–37	38.019719	-107 335334	82.68	0	11,598	34.2	0.0	24.1
09124500	1937 to current year	38 298883	-107 230056	339.34	0	10,886	28.0	0.5	14.9
09124700	1963–68	38.452214	-107 348112	3,456.67	1,046,900	9,919	21.4	0.6	4.4
09125000	1945–72	38.487767	-107.415057	35.02	0	9,671	21.8	0.0	0.1
09126000	1954 to current year	38 257213	-107.546726	66.88	13,520	10,862	32.7	0.0	20.2
09126500	1902–05, 1962–67	38.441099	-107.554225	210.95	13,520	9,764	25.4	0.4	7.4
09127000	1942–52	38.446377	-107.555614	230.46	13,520	9,631	24.6	0.5	6.8
09127500	1916–19, 1945–54, 1960–69	38 551933	-107 506169	42.27	0	9,623	20.3	0.0	0.4
09128000	1903 to current year	38 529153	-107.648947	3,970.10	1,202,856	9,872	21.5	0.6	4.3
09128500	1935–94	38.727768	-107 506723	43.09	0	9,164	25.7	0.0	4.8
09129000	1954–60	38.710544	-107 576169	63.69	0	8,920	24.8	0.1	3.7
09129600	1976–87	38.707486	-107.710336	166.86	23,395	8,096	19.5	1.6	1.6
09129800	1965–73	39 143318	-107.431165	38.37	0	9,300	33.3	0.0	0.3
09130500	1934–53	39.013322	-107 358385	132.97	0	8,692	28.6	0.6	2.0
09130600	1955–65	39 130817	-107 575334	7.40	0	9,423	29.7	0.0	0.5
09130800	1968–74	39 115540	-107 524500	27.57	0	9,275	29.2	0.7	0.1
09131100	1968–82	39 104150	-107 584501	11.90	5,990	10,208	33.9	0.0	2.7
09131200	1961–73	39.089708	-107 505333	49.82	5,990	9,403	29.9	0.6	0.7
09131500	1949–55	38 939990	-107.358108	257.13	26,940	8,651	27.3	0.5	1.7
09132000	1938–43, 1954–58	38.863048	-107.163939	21.12	0	10,323	36.3	0.8	7.7
09132050	1977–81	38 953880	-107 273662	95.25	0	9,631	32.9	0.4	9.4
09132500	1933 to current year	38 925823	-107.434221	526.23	26,940	8,888	28.2	0.5	4.5
09132700	1960–68	39.056928	-107.629502	1.35	0	10,320	34.2	0.0	0.3
09132800	1960–68	39.049428	-107.620891	1.43	0	10,438	34.5	0.0	0.1

Area covered by deciduous forest (percent)	Area covered by evergreen forest (percent)	Area covered by mixed forest (percent)	Area covered by shrubs young or stunted trees (percent)	Area covered by grass or herbaceous land (percent)	Sedimentary, clastic lithology, Mesozoic (percent)	Sedimentary, clastic (continental) lithology, Tertiary (percent)	Igneous or metamorphic lithologies (percent)	Sedimentary, mixed (continental and marine) lithology (percent)	Sedimentary, carbonate (marine) lithology (percent)	Population density (people per square mile)	Road density (miles per square mile)
1.8	64.0	0.0	0.1	20.5	0.0	0.0	83.9	5.5	10.5	0.12	1.36
1.5	54.9	0.0	0.3	24.5	0.0	1.1	87.8	4.6	6.5	0.12	0.83
4.7	57.0	0.5	2.6	21.7	1.8	0.6	79.1	2.6	15.9	0.23	0.77
13.2	26.1	0.5	0.4	35.3	37.3	1.0	27.2	0.0	34.5	0.89	0.65
6.1	69.3	0.0	0.0	18.0	0.0	30.0	70.0	0.0	0.0	2.16	2.19
18.5	35.6	0.4	1.3	28.4	65.9	9.2	24.8	0.0	0.0	3.26	2.11
13.3	42.0	1.5	0.7	33.3	2.4	0.0	10.7	0.0	87.0	0.67	0.77
17.3	31.9	0.9	2.4	29.9	42.0	4.0	24.7	1.8	27.5	1.91	1.41
20.0	28.8	1.0	7.8	26.4	45.1	6.0	23.3	2.4	23.2	1.71	1.41
16.0	50.8	1.1	0.0	16.0	0.0	48.4	51.6	0.0	0.0	0.89	0.59
17.1	50.3	1.8	0.1	15.9	0.0	52.9	47.1	0.0	0.0	0.89	0.62
24.2	46.0	2.1	1.1	15.9	6.1	55.3	38.7	0.0	0.0	0.78	0.84
32.9	33.6	3.5	6.6	14.5	35.6	31.7	32.7	0.0	0.0	0.79	0.98
28.3	27.3	2.8	17.8	11.4	30.0	38.9	28.8	2.2	0.0	0.79	1.13
27.9	26.9	2.8	18.3	11.2	29.5	39.2	29.0	2.2	0.0	0.79	1.19
13.9	40.4	1.2	10.9	19.9	20.5	9.2	53.6	2.6	14.1	0.90	1.14
7.0	67.6	1.9	5.3	12.6	4.6	0.0	86.6	0.0	8.8	0.08	0.86
7.2	67.5	2.0	6.6	11.4	3.5	0.1	89.7	0.0	6.7	0.08	0.73
7.6	49.6	1.8	26.9	7.8	20.4	8.5	68.0	0.0	3.1	0.08	1.05
6.6	62.1	1.4	3.8	17.4	0.8	0.0	75.9	0.0	23.3	0.13	1.60
11.1	41.1	0.7	6.0	35.2	1.9	7.3	89.1	1.6	0.0	0.04	0.94
11.0	39.2	1.0	10.1	33.1	4.2	7.2	87.0	1.5	0.0	0.04	0.99
8.8	40.6	1.3	26.3	16.6	13.4	6.9	75.6	0.5	3.6	0.51	1.21
4.4	43.3	0.0	0.0	40.1	0.0	0.0	99.2	0.0	0.0	0.12	0.35
10.2	47.6	0.3	1.9	30.9	0.3	5.9	93.0	0.7	0.0	0.09	0.48
10.9	46.7	0.5	7.5	26.1	1.4	6.0	92.0	0.5	0.0	0.08	0.65
21.8	51.7	0.9	4.1	17.7	14.0	6.5	79.6	0.0	0.0	0.04	0.21
4.7	18.8	0.1	0.0	40.8	0.0	3.9	96.1	0.0	0.0	0.12	0.37
12.4	33.5	0.3	0.2	29.6	0.0	12.3	87.1	0.6	0.0	0.12	0.76
8.9	28.6	0.1	0.2	37.0	0.0	0.2	99.8	0.0	0.0	0.12	0.47
11.4	39.6	0.5	4.4	26.8	0.0	7.5	91.2	1.3	0.0	0.10	0.76
12.1	37.9	1.2	22.0	17.5	11.7	7.5	74.4	1.1	5.2	0.59	1.13
26.9	52.7	1.1	3.9	14.6	0.0	10.1	89.9	0.0	0.0	0.04	0.55
12.0	46.4	2.9	0.0	16.9	0.0	19.7	80.3	0.0	0.0	0.08	0.50
24.4	40.5	2.6	9.6	12.0	9.6	38.1	51.5	0.8	0.0	0.60	1.12
25.2	37.9	2.4	12.8	11.4	11.7	40.3	47.3	0.7	0.0	0.74	1.18
43.1	31.8	1.6	6.9	15.5	20.7	13.4	65.8	0.1	0.0	0.08	0.92
14.4	37.8	1.3	20.6	16.9	11.3	9.3	73.8	1.0	4.6	0.58	1.15
52.9	15.6	4.8	6.3	14.9	39.2	13.0	47.8	0.0	0.0	0.14	0.33
52.6	16.0	4.0	8.1	13.1	44.1	13.4	42.4	0.0	0.0	0.46	0.42
36.2	22.0	2.1	18.9	5.4	64.2	16.3	18.9	0.5	0.0	0.90	1.52
65.0	19.4	2.0	1.8	10.8	19.5	78.7	0.0	1.8	0.0	0.14	0.18
61.8	12.6	3.1	9.4	6.8	11.0	79.0	6.6	3.3	0.0	0.15	0.64
36.8	32.3	10.9	0.3	16.0	0.0	86.2	2.7	11.1	0.0	0.58	0.26
57.9	17.7	5.4	1.1	14.9	0.0	76.5	20.5	3.0	0.0	0.58	0.39
5.6	74.4	0.5	0.0	13.5	0.0	36.1	63.9	0.0	0.0	0.58	0.34
43.9	31.9	5.5	1.3	12.9	0.0	64.3	34.0	1.7	0.0	0.58	0.44
59.0	14.6	3.5	10.5	6.6	7.5	79.2	11.3	2.1	0.0	0.24	0.61
18.2	53.6	0.9	0.1	15.0	15.5	52.2	32.3	0.0	0.0	0.65	1.52
40.1	29.3	2.1	3.3	12.9	30.7	39.8	29.5	0.0	0.0	0.24	0.54
53.3	19.7	3.3	8.1	8.1	21.7	57.4	19.9	1.0	0.0	0.20	0.54
5.9	86.2	0.4	0.0	5.9	0.0	59.9	40.1	0.0	0.0	0.58	0.00
1.7	92.6	0.2	0.1	3.1	0.0	59.9	40.1	0.0	0.0	0.58	0.00

Site number	Period of record	Latitude (decimal degrees)	Longitude (decimal degrees)	Drainage area (square miles)	Amount of upstream reservoir storage (acre-feet)	Mean basin elevation (feet)	Mean basin average annual precipitation (inches)	Area covered by developed land (percent)	Area covered by barren land (percent)
09132900	1960–73	39.032207	-107.613668	2.74	0	10,352	34.2	0.0	3.9
09132920	1968–74	39.044708	-107.566723	20.69	0	9,661	30.9	0.0	0.6
09132940	2001 to current year	38.982487	-107.531723	48.63	0	8,932	27.1	0.0	0.2
09132960	2001 to current year	38.925544	-107.518389	57.98	0	8,682	25.7	0.0	0.2
09132985	2001 to current year	38.964709	-107.567001	4.82	0	9,035	26.5	0.0	0.0
09132995	2001 to current year	38.903876	-107.562001	29.57	0	8,768	25.5	0.0	0.1
09133000	1921–32	38.899154	-107.563668	653.73	26,940	8,772	27.4	0.4	3.6
09134000	1936–47, 1985 to current year	38.869989	-107.504223	40.77	0	8,482	25.0	0.0	2.4
09134050	1976–79	38.874154	-107.588946	53.85	0	8,048	23.4	0.6	1.9
09134100	2000 to current year	38.857500	-107.621944	746.43	26,940	8,646	26.7	0.6	3.3
09134200	1976–79	38.806097	-107.687281	41.99	0	6,408	16.6	1.3	0.7
09134500	1936–56, 1960–69	38.926371	-107.793672	34.55	0	9,724	33.6	0.0	2.4
09134700	1960–69	38.926093	-107.792561	8.12	0	8,694	24.1	0.0	0.0
09135000	1917–26	38.881094	-107.785894	51.78	0	9,180	29.2	0.0	1.6
09135900	1976–96	38.798040	-107.732003	66.66	0	8,599	26.0	0.3	1.2
09135950	1997–2009	38.788318	-107.739781	923.70	26,940	8,383	25.4	0.8	2.8
09136100	2009 to current year	38.785167	-107.833417	969.58	26,940	8,264	24.8	0.9	2.7
09136200	1962–85	38.783040	-107.837840	5,247.56	1,253,191	9,444	21.9	0.7	3.9
09136500	1948–54	38.849983	-107.887841	42.61	0	7,515	18.5	1.2	0.2
09137050	1976–87	38.784706	-107.938954	57.09	0	7,061	16.6	1.6	0.5
09137800	1957–69	38.944704	-108.028125	11.10	0	9,586	32.1	0.0	0.7
09139200	1957–69	38.983592	-107.972012	14.39	0	10,143	38.2	1.9	3.7
09139500	1939–46	38.926926	-107.977289	22.23	0	9,330	31.7	1.3	2.4
09140200	1957–69	38.986647	-107.943677	6.18	0	9,738	36.1	1.2	0.9
09140500	1939–46	38.929426	-107.952566	10.36	0	8,736	28.3	2.6	0.6
09140700	1957–68	38.927204	-107.950622	10.36	0	8,736	28.3	2.6	0.6
09141000	1939–46	38.929982	-107.953400	4.13	0	8,203	22.5	0.8	0.1
09141200	1957–69	38.958037	-107.918676	10.31	0	9,314	32.2	0.0	0.1
09141500	1939–46	38.922759	-107.947566	13.32	0	8,794	28.5	0.5	1.9
09142000	1944–52	38.894426	-107.971455	55.78	0	8,753	28.0	1.7	1.1
09143000	1939 to current year	38.984703	-107.854508	27.30	0	9,878	38.3	0.0	3.7
09143500	1916 to current year	38.901649	-107.921176	42.07	0	9,347	33.2	0.6	2.4
09144000	1939–51	38.833594	-107.971733	47.66	0	8,987	30.9	2.0	2.1
09144200	1957–68, 1976–87	38.787761	-107.995344	201.07	0	8,046	24.0	2.0	1.3
09144250	1976 to current year	38.753039	-108.078403	5,636.13	1,253,191	9,279	21.6	0.8	3.7
09144500	1947–55	37.962772	-107.662838	18.08	0	11,396	40.0	2.5	19.7
09145000	1908, 1910–24	38.019160	-107.676171	41.92	0	11,362	38.8	2.5	20.7
09145500	1910–15	38.019716	-107.676171	27.90	0	11,353	38.2	0.1	40.4
09146000	1913–29	38.031105	-107.675060	75.07	0	11,267	38.1	2.0	28.6
09146020	2001 to current year	38.043327	-107.683115	76.85	0	11,224	37.8	2.0	28.0
09146200	1958 to current year	38.183879	-107.745892	149.25	0	9,963	30.9	2.2	15.8
09146400	1955–70	38.073603	-107.851174	14.08	0	10,217	30.6	0.0	17.2
09146500	1947–53, 1960–70	38.093325	-107.813673	16.50	0	10,942	35.2	0.0	38.2
09146550	1960–68	38.116380	-107.818117	12.21	0	9,369	27.3	0.1	5.5
09146600	1955–67	38.145546	-107.919787	8.28	0	9,057	25.6	0.0	0.0
09147000	1922–27, 1955–71, 1979 to current year	38.177768	-107.758393	97.30	0	9,165	26.1	1.0	9.7
09147025	1988 to current year	38.238045	-107.759226	264.55	0	9,492	28.2	1.8	12.4
09147100	1955–73	38.149436	-107.644782	47.46	0	10,656	33.3	0.0	17.0
09147500	1903–05, 1906, 1912 to current year	38.331377	-107.779504	447.58	0	9,230	26.4	1.2	9.3
09149400	1977–81	38.523875	-107.971177	76.79	0	7,583	17.8	1.5	0.1
09149420	1977–81	38.392210	-107.945065	41.73	0	8,461	22.2	0.0	0.1
09149480	1996–98	38.645818	-108.048958	202.54	0	7,070	14.3	1.6	0.2
09149500	1903–31, 1938 to current year	38.741928	-108.080903	1,119.50	0	7,838	18.9	2.7	4.7

Area covered by deciduous forest (percent)	Area covered by evergreen forest (percent)	Area covered by mixed forest (percent)	Area covered by shrubs young or stunted trees (percent)	Area covered by grass or herbaceous land (percent)	Sedimentary, clastic lithology, Mesozoic (percent)	Sedimentary, clastic (continental) lithology, Tertiary (percent)	Igneous or metamorphic lithologies (percent)	Sedimentary, mixed (continental and marine) lithology (percent)	Sedimentary, carbonate (marine) lithology (percent)	Population density (people per square mile)	Road density (miles per square mile)
1.8	91.4	0.0	0.1	2.3	0.0	58.9	41.1	0.0	0.0	0.58	0.00
32.7	53.5	4.7	0.1	7.8	0.0	56.4	37.1	6.5	0.0	0.58	0.30
62.4	27.1	3.4	2.0	3.9	2.4	75.2	19.5	2.8	0.0	0.53	0.45
62.3	24.4	3.0	5.7	3.3	15.8	65.4	16.5	2.3	0.0	0.54	0.47
71.5	17.2	1.1	3.1	6.7	18.1	81.9	0.0	0.0	0.0	0.58	0.29
70.1	15.8	3.5	5.0	4.7	41.2	58.8	0.0	0.0	0.0	0.58	0.21
55.5	19.2	3.1	8.8	7.1	25.1	56.4	17.5	1.0	0.0	0.29	0.54
59.0	12.6	4.4	16.3	4.2	66.9	13.8	19.3	0.0	0.0	0.49	0.76
47.0	13.5	3.4	27.5	3.1	70.9	13.3	15.2	0.7	0.0	1.41	0.91
54.4	18.9	3.0	10.5	6.5	31.4	51.1	16.4	1.1	0.0	0.62	0.66
3.5	26.3	0.0	53.3	0.4	85.9	9.8	2.2	2.1	0.0	3.31	1.02
36.0	34.6	1.9	4.5	18.1	9.9	38.5	45.8	5.7	0.0	1.31	0.34
81.6	1.5	2.4	7.0	3.1	45.6	32.9	0.0	21.5	0.0	1.19	0.75
41.7	28.1	1.7	8.6	13.2	27.5	30.9	30.5	11.1	0.0	1.29	0.57
33.4	32.7	1.3	11.9	10.4	41.8	24.0	23.7	10.5	0.0	1.82	0.83
47.1	20.7	2.5	14.3	6.2	37.6	44.4	15.4	2.6	0.0	1.10	0.77
44.9	20.3	2.4	15.5	5.9	39.1	42.3	14.6	3.9	0.0	1.22	0.82
20.7	34.1	1.5	20.2	14.1	19.9	15.4	59.7	1.5	3.5	0.74	1.11
33.4	26.8	0.1	27.8	1.5	70.1	18.4	0.3	11.3	0.0	1.93	0.75
25.0	23.0	0.1	34.9	1.1	67.8	14.2	0.2	17.8	0.0	2.32	1.12
49.9	30.1	4.0	1.3	9.6	6.0	73.0	20.9	0.0	0.0	1.14	0.21
17.2	54.8	3.2	0.4	10.3	0.0	14.8	85.2	0.0	0.0	1.73	1.53
30.0	44.9	2.4	4.0	6.9	12.7	23.4	57.7	6.3	0.0	1.54	1.66
38.5	38.3	2.3	0.6	8.5	0.0	10.2	89.8	0.0	0.0	2.15	1.91
34.1	34.9	1.4	8.1	5.3	11.6	13.6	53.8	21.0	0.0	2.07	2.29
34.1	34.9	1.4	8.1	5.3	11.6	13.6	53.8	21.0	0.0	2.07	2.29
45.5	24.9	1.7	13.1	3.7	13.5	7.2	47.5	31.7	0.0	1.67	2.04
51.5	32.3	2.0	5.0	5.1	7.3	17.7	73.4	1.6	0.0	2.28	0.74
43.3	32.2	1.5	8.1	3.9	10.0	13.7	58.0	18.3	0.0	2.28	1.24
34.1	37.1	1.8	8.0	5.0	16.5	16.0	51.9	15.7	0.0	1.82	1.83
34.5	46.3	1.4	0.4	10.1	0.0	30.3	69.3	0.4	0.0	2.50	0.47
44.5	35.5	1.2	4.0	7.4	4.1	23.6	61.4	10.9	0.0	3.70	1.17
39.4	35.3	1.1	4.6	6.5	8.1	20.9	54.2	16.9	0.0	5.85	1.71
29.3	30.8	1.2	17.9	5.4	35.7	22.5	29.7	12.1	0.0	3.92	1.52
20.6	33.2	1.4	21.2	13.3	22.3	15.5	56.7	2.3	3.2	1.05	1.15
11.7	27.9	0.8	0.1	35.3	0.0	0.0	97.9	0.0	2.1	1.09	1.32
9.7	28.3	0.8	0.4	35.8	0.0	0.0	92.4	6.0	1.7	0.93	1.04
7.8	29.1	1.1	0.3	19.4	3.5	0.0	88.9	0.0	7.6	1.09	0.74
8.9	29.6	0.9	0.5	27.9	1.8	0.0	88.4	3.7	6.1	1.00	1.07
9.0	30.4	0.9	0.6	27.4	2.2	0.0	87.4	3.6	6.8	1.00	1.08
22.2	33.1	1.5	2.9	16.3	20.0	8.1	58.6	6.6	6.7	1.25	1.61
22.4	48.4	1.9	0.2	9.9	22.6	31.6	45.8	0.0	0.0	1.08	0.68
17.7	32.0	3.1	0.1	8.1	3.0	26.3	70.7	0.0	0.0	1.10	0.38
37.6	46.7	4.7	0.3	4.4	31.6	39.7	28.8	0.0	0.0	1.08	0.88
26.8	9.8	0.0	1.2	62.1	100.0	0.0	0.0	0.0	0.0	0.92	1.93
28.3	35.9	1.5	2.7	14.9	55.1	20.6	24.3	0.0	0.0	1.04	1.39
23.2	36.0	1.4	3.7	14.7	37.6	12.8	42.1	3.7	3.8	1.15	1.65
14.8	39.8	2.5	0.6	24.5	2.2	2.9	92.5	0.0	2.4	1.08	0.15
26.0	39.0	1.4	4.7	12.6	38.4	18.8	37.4	2.9	2.5	1.08	1.41
22.8	40.6	0.2	16.1	1.4	92.4	0.1	0.0	7.4	0.0	5.75	2.05
38.9	51.5	0.4	6.3	2.3	100.0	0.0	0.0	0.0	0.0	0.78	0.98
13.7	33.4	0.3	34.2	2.8	84.4	5.5	0.0	10.1	0.0	1.47	1.84
18.5	32.6	0.7	19.2	6.3	59.8	18.8	15.1	5.4	1.0	4.46	2.13

Site number	Period of record	Latitude (decimal degrees)	Longitude (decimal degrees)	Drainage area (square miles)	Amount of upstream reservoir storage (acre-feet)	Mean basin elevation (feet)	Mean basin average annual precipitation (inches)	Area covered by developed land (percent)	Area covered by barren land (percent)
09149900	1980–81	38.479710	-108 313414	8.57	0	8,856	31.2	0.3	0.0
09149910	1980–81	38.614986	-108 206742	26.38	0	7,688	19.1	0.1	1.5
09150500	1938–54, 1976–83	38.734984	-108 161740	248.74	0	7,272	17.1	1.0	1.0
09151500	1922–23, 1970–89	38.756651	-108 260077	209.59	0	7,678	17.6	0.0	1.2
09152000	1917–82	38 961649	-108 230358	58.46	0	9,430	31.5	0.1	2.2
09152500	1896–99, 1901–06, 1916 to current year	38 983316	-108.450645	7,920.38	1,258,943	8,753	20.4	1.1	3.5
09152520	2000–2003	38 989149	-108.448700	7.30	0	4,809	9.3	3.7	10.2
09152900	1973–83	39 136925	-108.697320	15.62	0	5,230	10.8	2.3	1.7
09153000	1907–23	39 137481	-108.730933	17,049.88	2,538,886	8,719	22.1	1.4	3.9
09153270	1973–77	39 163592	-108.750933	170.92	0	5,994	13.5	0.3	2.4
09153290	1975–2000	39 211369	-108.803713	18.18	0	4,701	9.4	3.3	0.2
09153300	1973–83	39 183592	-108.787324	29.53	0	4,666	9.4	3.3	0.3
09153330	1979–82	39 396366	-108 981497	95.76	0	6,866	16.5	0.7	1.6
09153400	1973–83	39 308590	-108 983720	167.99	0	6,525	15.6	0.6	1.6
09160000	1976–82	39 331090	-108 930940	1.82	0	5,040	11.5	0.0	0.2
09160500	1976–82	39 323312	-108 930662	2.11	0	5,024	11.4	0.0	0.2
09161000	1976–82	39 329423	-108 943163	0.21	0	5,037	11.2	0.0	0.0
09163050	1973–82	39 293313	-108 933718	6.61	0	4,933	10.9	0.1	0.1
09163310	1973–82	39 297201	-108.866771	196.78	0	6,350	15.1	0.5	1.9
09163340	1973–82	39 265813	-108.842881	15.56	0	4,980	9.9	0.8	0.1
09163490	1973–83	39 221647	-108.892883	436.42	0	6,160	14.4	0.9	1.6
09163500	1951 to current year	39 132760	-109.027055	17,848.75	2,538,886	8,591	21.7	1.4	3.8
09163570	1983–88	38.850817	-108.782879	0.71	0	9,332	24.0	0.0	0.0
09163675	1983–86	39.081652	-109 217615	171.19	0	6,207	13.7	0.1	1.3
09165000	1951–96, 1998 to current year	37.638884	-108.060352	105.53	0	10,631	35.9	0.6	6.3
09166000	1941–44	37.639716	-108 327029	162.47	21,711	9,739	31.0	0.0	4.1
09166500	1895–1903, 1910–12, 1921 to current year	37.472493	-108.497591	504.55	21,711	9,695	31.2	0.4	3.1
09166950	1984 to current year	37.446105	-108.469256	69.28	0	8,602	26.4	0.1	0.0
09167000	1922–27, 1941–48	37.461382	-108 501480	71.54	0	8,558	26.2	0.2	0.0
09167450	1982–83	37 599159	-108.496202	83.88	0	8,146	23.8	0.0	0.0
09167500	1938–52	37 576936	-108 572594	815.20	21,711	9,105	28.3	0.3	1.9
09168100	1957–86	37.876658	-108 583150	147.23	0	7,939	21.4	0.0	0.7
09168500	1953–56	37 912769	-108.650096	168.19	0	7,812	20.8	0.0	0.6
09168730	1997–2003, 2008 to current year	38.044437	-108.905382	1,430.17	21,711	8,342	24.1	0.3	1.2
09168800	1979–81	38 113603	-108.858713	44.15	0	6,245	14.1	0.5	1.9
09169500	1917–22, 1971 to current year	38 310268	-108.885381	2,020.43	21,711	7,863	21.9	0.3	1.4
09170500	1944–52	38 383322	-108 996773	24.04	0	7,743	23.9	0.0	0.0
09170800	1971–73	38 331656	-108 900381	54.61	0	6,663	19.4	1.7	0.7
09171000	1944–52	38 329990	-108.871769	55.91	0	6,625	19.2	1.9	0.7
09171100	1971 to current year	38 356934	-108.833435	2,142.65	21,711	7,763	21.6	0.4	1.4
09171200	1959–65	37 948049	-107.877009	42.76	0	11,173	36.8	2.5	27.3
09171500	1895–99, 1910	37 992770	-108.023957	175.95	0	10,482	33.1	1.3	18.8
09172000	1941–59	37 958326	-108.005901	1.08	0	9,247	25.0	0.0	0.0
09172100	1955–63	38 101658	-107 923398	9.52	0	9,658	27.2	0.8	5.9
09172500	1909–12, 1930–34, 1942 to current year	38.042491	-108 132296	309.74	0	9,944	30.2	0.9	11.9
09172600	1976–80	37 923326	-108 131741	4.77	0	10,125	35.1	0.0	0.5
09172800	1976–80	37.889159	-108 197578	5.34	0	10,612	36.6	0.0	4.3
09173000	1941–61, 1962–67, 1975–81	37 969992	-108.195633	40.73	0	10,084	34.5	0.0	3.8
09173500	1942–51	38 203878	-108.057570	29.26	0	8,985	25.3	0.0	0.0
09174000	1953–62	38 263045	-108 397863	650.18	0	9,212	26.9	0.5	5.9
09174500	1942–51	38 273601	-108 362862	38.28	0	7,645	19.7	0.6	0.0
09174600	1995 to current year	38 244157	-108 502034	743.95	0	8,955	25.7	0.5	5.2

Area covered by deciduous forest (percent)	Area covered by evergreen forest (percent)	Area covered by mixed forest (percent)	Area covered by shrubs young or stunted trees (percent)	Area covered by grass or herbaceous land (percent)	Sedimentary, clastic lithology, Mesozoic (percent)	Sedimentary, clastic (continental) lithology, Tertiary (percent)	Igneous or metamorphic lithologies (percent)	Sedimentary, mixed (continental and marine) lithology (percent)	Sedimentary, carbonate (marine) lithology (percent)	Population density (people per square mile)	Road density (miles per square mile)
51.8	42.2	0.7	0.9	3.8	100.0	0.0	0.0	0.0	0.0	0.39	0.82
28.0	46.9	0.2	21.3	1.4	100.0	0.0	0.0	0.0	0.0	0.39	1.00
23.8	36.9	0.4	28.7	0.9	93.0	0.0	0.3	4.0	2.7	0.41	1.17
33.1	42.3	0.8	17.8	3.1	97.8	0.0	2.2	0.0	0.0	0.31	0.61
26.3	37.1	3.2	2.1	26.1	13.3	39.4	47.3	0.0	0.0	0.45	0.69
19.8	33.3	1.2	23.2	10.9	37.3	14.4	43.1	2.7	2.5	1.49	1.23
0.0	0.0	0.0	80.4	0.1	77.7	22.3	0.0	0.0	0.0	1.66	2.10
0.4	9.5	0.0	62.4	0.0	91.7	0.1	0.0	8.2	0.0	4.15	1.72
18.6	35.0	1.2	22.7	10.3	27.7	22.7	35.6	2.6	11.3	2.81	1.40
6.8	28.5	0.0	55.0	0.4	75.1	24.6	0.0	0.3	0.0	1.68	1.11
0.0	0.0	0.0	37.3	0.3	100.0	0.0	0.0	0.0	0.0	1.45	3.02
0.0	0.0	0.0	35.8	0.2	99.1	0.2	0.0	0.7	0.0	1.49	3.21
17.0	47.1	0.0	31.5	1.1	76.7	23.3	0.0	0.0	0.0	0.15	0.85
12.2	41.0	0.0	43.3	0.7	75.3	24.7	0.0	0.0	0.0	0.28	0.86
0.0	0.0	0.0	99.8	0.0	100.0	0.0	0.0	0.0	0.0	1.43	1.54
0.0	0.0	0.0	99.8	0.0	100.0	0.0	0.0	0.0	0.0	1.43	1.48
0.0	0.0	0.0	100.0	0.0	100.0	0.0	0.0	0.0	0.0	1.43	3.94
0.0	0.0	0.0	99.5	0.0	100.0	0.0	0.0	0.0	0.0	1.43	2.11
13.7	36.4	0.0	45.6	0.6	73.6	26.4	0.0	0.0	0.0	0.37	0.78
0.0	0.7	0.0	94.8	0.1	100.0	0.0	0.0	0.0	0.0	1.43	1.78
10.9	32.3	0.0	49.6	0.6	76.4	23.6	0.0	0.0	0.0	0.51	1.02
18.1	34.7	1.2	24.0	9.8	30.0	22.6	34.0	2.6	10.8	2.77	1.39
21.8	43.4	34.5	0.2	0.1	100.0	0.0	0.0	0.0	0.0	0.19	0.21
0.4	39.1	0.0	57.7	0.7	41.7	58.3	0.0	0.0	0.0	0.08	0.43
17.4	55.8	2.9	0.1	15.8	46.2	4.7	13.0	0.5	35.6	0.07	0.85
44.6	32.3	5.2	0.6	11.8	87.5	0.3	8.8	0.0	3.4	0.04	1.04
38.1	42.4	3.9	0.7	9.9	79.1	1.1	5.9	0.1	13.8	0.28	0.97
36.4	53.9	4.6	1.7	1.7	99.7	0.1	0.2	0.0	0.0	0.86	1.57
35.4	54.6	4.5	2.0	1.7	99.7	0.1	0.2	0.0	0.0	0.92	1.62
37.5	21.8	0.1	33.1	4.5	98.4	1.6	0.0	0.0	0.0	0.06	1.10
36.2	42.1	2.8	6.3	7.7	86.4	1.3	3.7	0.1	8.6	0.35	1.12
37.6	35.8	0.4	18.8	5.0	97.8	1.2	1.0	0.0	0.0	0.06	1.03
34.1	37.8	0.3	21.1	4.4	98.1	1.0	0.9	0.0	0.0	0.06	1.04
29.3	42.9	1.7	17.0	5.2	91.3	1.4	2.3	0.0	5.0	0.27	1.28
0.0	21.3	0.0	74.7	0.5	57.5	17.7	0.0	0.0	24.8	0.23	2.08
22.0	43.5	1.2	25.3	3.9	91.1	2.3	1.7	0.6	4.3	0.24	1.52
12.8	74.2	0.0	10.8	1.1	100.0	0.0	0.0	0.0	0.0	0.18	2.00
6.1	55.0	0.0	26.1	0.6	75.3	20.8	0.0	1.1	2.8	0.23	2.36
6.0	53.7	0.0	26.3	0.6	73.5	22.6	0.0	1.1	2.7	0.23	2.40
20.9	43.2	1.1	26.6	3.7	89.1	3.9	1.6	0.9	4.4	0.24	1.55
15.1	26.1	3.0	0.2	23.1	19.7	9.4	65.3	0.0	5.6	8.36	2.06
24.9	28.1	4.4	0.9	20.2	45.6	9.3	41.9	0.0	3.2	3.92	1.77
32.9	1.2	0.0	0.4	65.1	100.0	0.0	0.0	0.0	0.0	0.54	2.10
33.1	26.5	1.7	0.2	31.5	57.1	12.7	30.2	0.0	0.0	0.57	1.72
29.9	30.3	4.4	1.6	20.1	63.4	6.5	26.1	0.0	4.0	2.46	1.69
47.8	29.4	9.1	0.0	13.1	98.7	0.0	1.3	0.0	0.0	0.53	0.48
8.7	65.0	6.5	0.0	15.3	67.1	0.0	32.9	0.0	0.0	0.23	2.07
32.3	38.7	10.7	0.1	14.3	79.8	0.0	20.2	0.0	0.0	0.34	1.24
54.0	10.3	0.0	2.3	32.7	90.2	9.8	0.0	0.0	0.0	0.50	1.75
33.8	36.5	3.1	3.3	15.4	80.4	3.5	13.8	0.0	2.3	1.39	1.47
46.4	37.1	0.2	7.1	1.4	98.0	2.0	0.0	0.0	0.0	0.50	1.25
33.2	38.7	2.7	4.3	13.5	82.3	3.7	12.1	0.0	2.0	1.27	1.43

Site number	Period of record	Latitude (decimal degrees)	Longitude (decimal degrees)	Drainage area (square miles)	Amount of upstream reservoir storage (acre-feet)	Mean basin elevation (feet)	Mean basin average annual precipitation (inches)	Area covered by developed land (percent)	Area covered by barren land (percent)
09174700	1976–80	37.910826	-108 336195	7.26	0	9,590	30.6	0.0	2.8
09175000	1940–52, 1975–80	37 975825	-108 327861	56.46	6,851	8,642	25.5	0.5	1.2
09175400	1976–80	38 175546	-108 331750	41.39	10,039	7,596	18.5	2.8	0.0
09175500	1917–29, 1940–81	38 217768	-108 566481	1,067.63	16,890	8,483	23.6	0.8	3.7
09175900	1966–78	38.092214	-108.622039	78.85	0	7,257	17.9	0.9	0.0
09176500	1946–53	38 368878	-108 345639	18.47	0	8,958	31.5	0.8	0.0
09177000	1954–62, 1973–94, 1996 to current year	38 357212	-108.712875	1,501.03	16,890	8,017	21.9	0.7	2.7
09179000	1944–52	38.435266	-108 922882	75.07	0	8,684	28.4	0.0	1.5
09179200	1979–85	38 533043	-108 970940	31.83	0	6,487	18.5	0.0	1.1
09179500	1936–54	38.681373	-108 980387	4,345.63	38,601	7,750	21.3	0.5	1.8
09180000	1950 to current year	38.797208	-109 195114	4,570.59	38,601	7,688	21.1	0.5	1.8
09180500	1895 to current year	38.810541	-109 293449	23,951.03	2,577,487	8,237	21.0	1.2	3.3
09180920	1979–81	38.696930	-109 255115	6.68	0	6,090	15.3	0.0	0.1
09180970	1979–81	38.706374	-109 315116	17.66	0	5,800	14.2	0.0	2.3
09181000	1950–55	38.724984	-109 345117	20.51	0	5,635	13.7	0.0	2.0
09181500	1950–53	38.729150	-109 375673	33.24	0	6,003	14.4	0.1	1.7
09182000	1951–55, 1958–75	38 592764	-109 265671	8.71	0	9,382	30.6	0.0	3.0
09182200	1992–2001	38.612485	-109 332338	18.13	0	8,177	24.5	0.0	1.8
09182400	1992 to current year	38.673871	-109.450117	53.21	0	7,005	19.9	0.8	1.4
09182500	1950–55, 1957–58	38.679149	-109.449284	53.47	0	6,993	19.8	0.8	1.4
09182900	1959–66	38.648592	-109.599285	142.35	0	4,831	9.6	0.9	10.8
09183000	1950–55, 1957, 1966–89	38.612759	-109.579840	162.20	0	4,808	9.6	0.9	11.0
09183500	1954–59, 1987 to current year	38.483040	-109.404004	26.21	0	8,630	27.1	0.0	3.5
09183600	2003 to current year	38.485718	-109.411104	26.60	0	8,586	26.9	0.0	3.6
09184000	1949–71, 1972–93	38 562205	-109 514006	74.49	0	7,195	19.8	0.0	3.9
09184500	1955–59	38.436097	-109 354837	15.68	0	8,669	25.7	0.0	2.0
09185000	1955–59	38 540261	-109 500672	57.84	0	6,899	18.7	3.1	2.2
09185500	1951–71	38 243320	-109.440114	371.42	0	6,560	14.1	1.5	2.2
09185800	1958–80	37.841105	-109 505400	2.18	0	10,095	34.0	0.0	1.2
09186000	1950–57	37.844438	-109 518734	4.66	0	9,868	32.7	0.0	0.7
09186500	1949–71, 1988–91	37 972214	-109 519289	30.59	0	8,646	26.4	0.0	1.3
09187000	1950–57	38.062489	-109 574288	115.00	0	7,325	16.7	0.6	1.8
09187500	1949–57	38 151655	-109.625675	261.12	0	6,902	15.6	0.4	4.7
09187550	1983–88	38 151655	-109.625675	261.12	0	6,902	15.6	0.4	4.7
09188500	1931–1992, 1993 to current year	43.018829	-110.118220	466.97	0	9,295	26.3	0.1	2.6
09189000	1938–54	43.005495	-110.142388	147.50	0	8,048	20.1	0 3	0.0
09189495	1982–84	42.946325	-110 395174	42.69	0	8,830	29.6	0.0	0.0
09189500	1954–74	42 943611	-110 385278	43.50	0	8,815	29.4	0.0	0.0
09189550	1982–85	42 907778	-110 333889	32.98	0	8,472	25.9	0.0	0.0
09190000	1931–54, 1982–85	42 928273	-110 198500	106.14	0	8,383	24.2	0.0	0.0
09190500	1913–18	42.853830	-110.071827	170.03	0	8,073	20.5	0.1	0.0
09191000	1912–32	42.783276	-109 967378	1.60	0	7,315	11.9	0.0	0.0
09191300	1982–84	42.765000	-110 522500	21.48	0	9,363	41.0	0.0	0.6
09191500	1938–54	42.774940	-110 153496	200.40	0	8,167	23.4	0.0	0.2
09192500	1938–40	42.672165	-110.006267	236.60	0	8,024	21.6	0.1	0.2
09193000	1938–72	43.078554	-109 995439	35.53	17,500	9,545	29.5	0.0	1.5
09194500	1938–41	43.002721	-109 939602	40.02	0	8,239	20.2	0.0	0.0
09195500	1938–41	42 933275	-110.000714	9.32	0	7,516	13.8	1.2	0.0
09196000	1938–44	42.833277	-109.850708	110.22	0	9,583	30.1	1.0	3.1
09196500	1954–97, 2000 to current year	43.030502	-109.770153	75.72	0	10,425	36.5	0.0	4.5
09197000	1910–12, 1915–18, 1985–86, 1988–	42.894944	-109.843764	100.81	0	9,796	31.9	0.1	3.4
09197500	1904–06	42.883277	-109.867376	103.79	0	9,727	31.3	0.1	3.3

Area covered by deciduous forest (percent)	Area covered by evergreen forest (percent)	Area covered by mixed forest (percent)	Area covered by shrubs young or stunted trees (percent)	Area covered by grass or herbaceous land (percent)	Sedimentary, clastic lithology, Mesozoic (percent)	Sedimentary, clastic (continental) lithology, Tertiary (percent)	Igneous or metamorphic lithologies (percent)	Sedimentary, mixed (continental and marine) lithology (percent)	Sedimentary, carbonate (marine) lithology (percent)	Population density (people per square mile)	Road density (miles per square mile)
63.6	13.4	9.5	0.5	10.1	96.6	0.0	3.4	0.0	0.0	0.09	2.02
46.5	22.2	2.7	13.1	5.0	97.4	0.0	2.6	0.0	0.0	0.21	1.35
20.7	36.3	0.0	5.5	1.2	98.8	1.2	0.0	0.0	0.0	0.27	2.76
28.6	41.2	2.0	8.3	10.2	87.4	2.6	8.6	0.0	1.4	0.97	1.55
19.2	30.9	0.0	42.2	2.2	100.0	0.0	0.0	0.0	0.0	0.23	1.31
70.8	21.1	2.3	1.1	3.7	86.3	13.7	0.0	0.0	0.0	0.50	0.70
25.7	39.9	1.5	17.1	7.6	89.9	2.4	6.5	0.0	1.2	0.80	1.55
33.6	44.2	2.3	7.9	10.3	75.9	6.7	5.3	12.1	0.0	0.26	1.68
5.3	56.9	0.0	35.6	0.3	66.8	0.0	0.0	10.8	22.4	0.23	0.52
22.4	42.9	1.2	22.6	5.3	87.8	3.4	4.6	1.5	2.8	0.43	1.52
21.9	43.1	1.1	23.3	5.1	87.5	3.4	4.4	2.0	2.6	0.43	1.51
17.8	35.7	1.1	26.6	8.3	43.9	18.8	26.3	2.3	8.6	2.15	1.39
0.7	52.7	0.0	44.6	0.0	58.5	27.3	0.0	10.5	3.7	0.46	1.19
0.4	33.3	0.0	63.1	0.0	47.6	10.5	0.0	32.1	9.7	0.46	0.74
0.4	28.8	0.0	68.0	0.0	41.9	9.2	0.0	40.5	8.4	0.46	0.80
3.0	37.3	0.0	57.2	0.0	61.9	4.7	0.0	33.4	0.0	0.46	0.87
32.4	46.7	11.4	3.1	3.1	66.3	10.6	19.4	3.6	0.0	0.46	1.92
18.9	62.5	5.5	8.2	1.5	65.5	17.2	14.2	3.1	0.0	0.46	1.95
14.9	47.9	2.3	28.8	1.0	55.4	17.7	13.0	13.9	0.0	0.46	2.11
14.9	47.7	2.3	29.1	1.0	55.5	17.8	13.0	13.8	0.0	0.46	2.11
0.0	0.5	0.0	77.1	10.1	97.5	0.6	0.0	0.0	1.9	0.08	2.37
0.0	0.5	0.0	77.9	9.1	97.8	0.5	0.0	0.0	1.7	0.08	2.13
30.8	40.8	4.1	14.4	5.8	66.1	6.4	13.0	14.5	0.0	0.20	1.54
30.3	40.3	4.1	15.5	5.7	66.6	6.3	12.8	14.3	0.0	0.20	1.56
18.3	35.3	1.5	35.8	3.1	87.3	2.2	4.6	5.8	0.0	0.62	1.64
34.9	51.9	4.6	2.1	3.4	59.4	23.9	10.0	6.7	0.0	0.08	1.40
16.6	33.1	1.8	37.4	1.8	83.2	7.5	4.5	4.8	0.0	0.11	2.95
0.3	24.1	0.0	63.7	1.3	97.2	0.1	0.0	1.4	1.3	0.08	1.81
22.0	34.5	11.4	2.0	28.9	46.9	15.1	38.0	0.0	0.0	0.08	1.54
19.9	29.5	26.8	1.3	21.7	60.8	7.8	31.4	0.0	0.0	0.06	1.78
35.7	35.3	12.3	6.7	7.5	59.7	24.1	16.2	0.0	0.0	0.07	1.15
16.9	53.5	1.2	22.0	1.8	96.0	0.9	0.5	2.7	0.0	0.05	1.72
12.3	38.2	2.1	38.4	1.8	84.3	3.6	2.3	9.6	0.2	0.05	1.36
12.3	38.2	2.1	38.4	1.8	84.3	3.6	2.3	9.6	0.2	0.05	1.36
4.1	31.0	0.2	34.1	21.0	16.8	11.8	25.7	26.3	17.7	0.16	0.80
2.7	27.5	0.1	49.5	0.5	0.0	97.2	0.0	2.8	0.0	0.15	1.30
1.0	42.4	0.5	48.5	4.7	54.9	42.1	0.0	2.9	0.1	0.14	0.61
1.0	42.6	0.5	48.1	4.6	53.8	43.2	0.0	2.8	0.1	0.14	0.61
1.6	52.0	0.2	34.8	3.6	35.8	62.6	0.0	0.6	1.1	0.15	0.53
1.4	35.6	0.3	43.2	3.0	33.2	65.1	0.0	1.3	0.4	0.15	0.78
1.2	23.9	0.2	51.7	2.4	20.7	78.2	0.0	0.8	0.2	0.15	1.09
0.0	0.0	0.0	92.5	1.8	0.0	100.0	0.0	0.0	0.0	0.12	2.01
1.0	45.6	0.2	31.7	17.7	48.9	10.0	0.0	20.0	21.1	0.34	0.36
1.4	23.7	0.1	51.6	3.6	20.5	73.0	0.0	3.9	2.6	0.32	1.09
1.2	20.0	0.1	53.8	3.6	17.3	77.1	0.0	3.3	2.2	0.31	1.23
4.3	29.4	0.3	29.0	27.9	0.5	10.6	65.3	22.9	0.7	0.15	0.27
9.1	31.9	0.3	54.0	0.2	0.0	20.6	19.2	60.2	0.0	0.15	1.06
0.0	0.4	0.0	66.2	0.2	0.0	95.2	0.0	4.8	0.0	0.15	2.49
2.4	16.3	0.2	33.6	28.3	0.0	12.2	73.2	14.6	0.0	0.15	0.82
1.0	15.2	0.3	34.3	40.0	0.0	0.0	97.9	2.1	0.0	0.15	0.06
2.6	17.8	0.3	33.8	30.2	0.0	7.5	80.1	12.4	0.0	0.15	0.36
2.5	17.3	0.2	34.7	29.5	0.0	7.6	77.8	14.6	0.0	0.15	0.43

Site number	Period of record	Latitude (decimal degrees)	Longitude (decimal degrees)	Drainage area (square miles)	Amount of upstream reservoir storage (acre-feet)	Mean basin elevation (feet)	Mean basin average annual precipitation (inches)	Area covered by developed land (percent)	Area covered by barren land (percent)
09198000	1903–04, 1914–54	42.866610	-109.867376	104.40	0	9,713	31.2	0.2	3.3
09198500	1938–71	42.881389	-109.711944	82.49	0	9,634	31.1	0.0	1.6
09199000	1904–06	42.833278	-109.750704	41.15	0	9,243	28.0	0.1	0.5
09199500	1938–71	42.855779	-109.720703	39.04	0	9,353	28.9	0.0	0.5
09200000	1904–05	42.849945	-109.734037	39.53	0	9,327	28.7	0.0	0.5
09201000	1914–69	42.750222	-109.728758	535.52	40,130	8,678	23.2	0.8	1.0
09201500	1938–39	42.856337	-109.617366	1.20	0	7,677	15.1	0.0	0.0
09202000	1938–73	42.819390	-109.717092	136.45	22,280	9,634	30.7	0.0	1.5
09203000	1938–92	42.672500	-109.421667	81.96	0	9,763	30.3	0.0	1.5
09204000	1938–71	42.743611	-109 510833	44.78	0	9,642	29.9	0.0	0.1
09204500	1904–06, 1914–24, 1930–32	42.699945	-109.717367	0.11	0	6,932	9.8	24.4	0.0
09205000	1954 to current year	42 567166	-109 930151	1,252.85	62,410	8,403	20.7	0.6	0.7
09205490	1982–84	42.661046	-110.489619	29.74	0	9,440	41.7	0.0	1.7
09205500	1915–16, 1931–72	42.658271	-110 342391	63.28	0	8,953	35.0	0.0	0.9
09206000	1939–54	42.602714	-110.456284	32.26	0	9,208	37.7	0.0	1.2
09207500	1938–42	42 511051	-110 309055	50.79	0	8,798	32.0	0.0	0.8
09207700	1965–73	42 390219	-110 253219	67.96	0	8,361	23.3	0.3	0.2
09208000	1940–42, 1950–81	42 506944	-110.674722	6.59	0	9,003	36.6	0.0	0.0
09208400	1982–84	42 293827	-110.440724	122.08	0	8,600	28.6	0.0	0.1
09208500	1913–16, 1940–49	42 222163	-110 320165	167.09	0	8,419	25.9	0.0	0.1
09209400	1963 to current year	42 192722	-110.163214	3,824.58	67,621	8,139	19.7	0.5	1.0
09210000	1941–42	42 161050	-110.525726	60.90	0	8,379	26.5	0.0	0.1
09210500	1951 to current year	42.096053	-110.416556	153.77	0	8,112	22.8	0.0	0.1
09211000	1914–19, 1931–53	42.097167	-110 222939	210.66	0	7,828	19.5	0.0	0.9
09211200	1963 to current year	42.021060	-110.049876	4,197.74	412,921	8,077	19.3	0.5	1.2
09212500	1910–11, 1939–87	42 571339	-109 283461	93.89	0	9,523	28.2	0.0	1.3
09213500	1914–17, 1920–24, 1926–34, 1953 to current year	42 316895	-109.485687	322.71	0	7,862	15.9	0.5	0.4
09214000	1939–71	42 533841	-109 204847	20.25	0	9,767	30.2	0.0	2.7
09214500	1954–81	42 236621	-109 312904	140.58	0	7,696	14.9	0.0	0.4
09215000	1954–73	42 129678	-109 323738	388.81	0	7,256	11.0	0.2	0.3
09215550	1981–99	42.073289	-109.479298	1,138.83	47,200	7,404	12.6	0.5	0.3
09216000	1954–81	42.010233	-109 583190	1,419.54	47,200	7,279	11.7	0.5	0.6
09216050	1972–	41 947455	-109.688194	1,522.70	47,200	7,239	11.5	0.5	0.5
09216500	1891, 1894–1906, 1914–39	41 533298	-109.484024	7,484.47	460,121	7,589	15.2	0.5	1.2
09216545	1975–81	41.493018	-108 513716	305.57	0	7,183	8.2	0.2	0.5
09216562	1976–81	41.647741	-108.997896	973.63	0	7,011	8.2	0.7	0.6
09216565	1976–81	41.198850	-108 998453	34.69	0	7,770	13 7	0.0	0.0
09216576	1975–76	41 206906	-109.053177	35.90	0	7,597	12.9	0.0	0.0
09216578	1976–80	41 217184	-108 916228	3.75	0	7,603	12.5	0.0	0.0
09216750	1976–81	41.630519	-108 989007	522.63	0	7,262	11.1	0.2	0.2
09217000	1951 to current year	41 516354	-109.449023	9,733.40	460,121	7,455	13.9	0.6	1.3
09217900	1937–1939, 1966–86, 1992 to current year	40 959113	-110 580166	124.60	0	10,473	34.2	0.3	17.3
09218500	1939–98	41.031614	-110 579332	137.71	32,470	10,338	33.4	0.3	15.7
09219000	1913–24, 1937–55	41 349951	-110 334046	274.54	32,470	8,953	24.1	1.4	7.9
09220000	1939–99, 2001 to current year	41.054114	-110 398492	55.38	14,000	10,363	31.3	0.6	10.3
09220500	1939–81	41.022169	-110.479329	38.12	0	9,786	28.6	0.7	0.7
09221500	1941–57	41 266617	-110 334045	210.59	14,000	9,099	23.7	1.2	2.8
09222000	1937–57, 1962–83	41.452176	-110.172929	799.72	46,470	8,297	18.9	1.8	3.7
09222200	1980–81	41 560504	-110.659061	53.81	0	7,287	13.3	0.0	0.0
09222250	1980–81	41 582778	-110.568889	388.28	0	7,233	13.3	0.7	0.4
09222300	1976–80	41 581615	-110 562391	470.56	0	7,186	12.8	1.1	0.4
09222400	1975–81	41 538008	-110 229321	970.20	0	7,171	12.9	0.8	1.1

Area covered by deciduous forest (percent)	Area covered by evergreen forest (percent)	Area covered by mixed forest (percent)	Area covered by shrubs young or stunted trees (percent)	Area covered by grass or herbaceous land (percent)	Sedimentary, clastic lithology, Mesozoic (percent)	Sedimentary, clastic (continental) lithology, Tertiary (percent)	Igneous or metamorphic lithologies (percent)	Sedimentary, mixed (continental and marine) lithology (percent)	Sedimentary, carbonate (marine) lithology (percent)	Population density (people per square mile)	Road density (miles per square mile)
2.5	17.2	0.2	34.6	29.4	0.0	8.0	77.3	14.7	0.0	0.15	0.48
3.7	31.2	0.3	40.2	17.0	0.0	2.5	77.4	20.1	0.0	0.15	0.34
2.2	15.2	0.4	58.8	14.0	0.0	7.4	70.5	22.1	0.0	0.15	0.45
2.3	15.9	0.4	59.7	14.8	0.0	4.1	74.4	21.6	0.0	0.15	0.41
2.2	15.7	0.4	59.7	14.6	0.0	4.6	73.4	22.0	0.0	0.15	0.41
3.0	18.0	0.2	47.2	12.4	0.1	31.5	43.1	25.0	0.3	0.15	1.20
13.0	48.4	0.2	38.0	0.0	0.0	0.0	89.2	10.8	0.0	0.12	2.77
1.5	22.3	0.9	49.6	18.2	0.0	3.7	78.4	17.9	0.0	0.14	0.26
4.5	45.6	0.7	33.3	12.5	0.0	1.8	75.4	22.7	0.0	0.12	0.14
2.5	51.9	1.0	39.8	3.1	0.0	0.1	93.1	6.9	0.0	0.12	0.01
0.0	0.0	0.0	0.3	3.2	0.0	100.0	0.0	0.0	0.0	0.12	0.33
2.1	16.7	0.3	58.5	9.0	0.0	44.1	38.7	16.9	0.3	0.13	1.12
0.9	49.9	0.1	23.8	21.0	71.6	10.8	0.0	12.7	4.9	0.28	0.13
1.5	45.2	0.2	32.7	13.8	47.8	36.4	0.0	11.9	3.9	0.36	0.61
1.7	54.7	0.1	24.2	15.5	49.5	19.1	0.0	22.5	9.0	0.36	0.57
1.6	40.7	0.1	33.5	11.3	31.5	48.5	0.0	14.3	5.7	0.38	0.85
0.3	24.4	0.1	67.6	5.8	0.7	74.6	0.0	0.0	24.7	0.12	2.71
0.1	41.3	0.1	47.7	7.6	52.9	10.9	0.0	1.8	34.3	0.08	1.55
0.5	48.3	0.0	36.4	14.3	73.1	2.1	0.0	10.1	14.7	0.10	1.05
0.5	39.2	0.0	45.6	11.1	59.9	18.4	0.0	7.4	14.3	0.10	1.21
1.6	17.6	0.1	58.1	8.5	8.9	60.7	15.9	10.3	4.0	0.18	1.30
1.1	37.8	0.0	41.4	16.7	68.5	16.0	0.0	8.3	7.2	0.09	0.65
1.1	27.8	0.0	48.2	21.3	75.2	15.2	0.0	3.6	6.0	0.10	0.92
0.8	20.3	0.0	58.2	16.1	56.6	36.4	0.0	2.6	4.4	0.11	1.18
1.5	17.0	0.1	59.1	8.6	11.1	60.8	14.5	9.5	3.9	0.17	1.32
6.6	47.7	0.3	26.8	15.2	0.0	10.9	58.7	30.3	0.0	0.12	0.77
2.2	14.2	0.1	70.8	10.3	0.0	68.9	19.5	11.6	0.0	0.12	1.46
5.1	41.8	0.2	23.0	22.3	0.0	6.6	66.3	27.0	0.0	0.12	0.71
0.8	6.6	0.1	82.4	6.8	0.0	78.3	15.6	4.8	0.0	0.12	1.67
0.0	0.1	0.0	91.4	7.4	0.3	99.7	0.1	0.0	0.0	0.05	1.58
0.7	4.9	0.0	81.3	9.4	0.1	88.2	7.6	3.9	0.0	0.08	1.73
0.6	3.9	0.0	80.2	10.8	0.1	90.6	6.1	3.1	0.0	0.08	1.80
0.6	3.7	0.0	80.1	11.5	0.1	91.2	5.7	2.9	0.0	0.08	1.85
0.9	10.3	0.1	70.6	8.8	6.6	76.0	9.2	5.9	2.2	0.16	1.55
0.0	0.1	0.0	96.4	2.8	0.2	99.8	0.0	0.0	0.0	0.04	1.07
0.0	0.1	0.0	94.8	3.6	32.1	67.9	0.0	0.0	0.0	0.04	1.51
0.0	1.5	0.0	94.2	3.8	21.0	79.0	0.0	0.0	0.0	0.04	1.35
0.1	0.5	0.0	87.1	9.9	37.6	62.4	0.0	0.0	0.0	0.04	1.45
0.0	0.0	0.0	95.4	4.6	100.0	0.0	0.0	0.0	0.0	0.04	1.08
0.0	0.4	0.0	91.9	6.7	50.9	49.1	0.0	0.0	0.0	0.04	1.24
0.7	8.0	0.1	75.4	8.0	14.8	71.8	7.1	4.5	1.7	0.52	1.56
0.7	60.3	0.8	12.9	6.6	2.9	8.0	0.0	82.8	6.3	0.12	0.55
1.6	60.3	2.1	12.2	6.1	2.9	10.2	0.0	81.2	5.7	0.12	0.61
2.9	35.8	2.2	27.6	6.8	4.5	45.2	0.0	47.5	2.9	0.19	1.28
0.4	65.6	0.3	11.6	9.0	0.9	17.0	0.0	81.2	0.9	0.12	0.54
2.3	79.2	2.1	7.6	6.2	15.4	42.5	0.0	38.0	4.0	0.13	1.98
3.1	46.9	1.6	21.5	6.0	13.7	56.6	0.0	28.8	1.0	0.73	1.32
2.3	27.4	1.2	41.6	5.4	7.8	67.0	0.0	23.9	1.2	1.01	1.46
2.9	0.5	0.6	63.0	32.5	11.1	88.9	0.0	0.0	0.0	0.27	1.80
2.4	0.9	0.9	67.7	25.9	34.5	65.4	0.0	0.0	0.1	0.22	1.77
2.0	0.7	0.8	71.3	22.6	44.7	55.2	0.0	0.0	0.1	0.36	1.87
2.0	3.4	0.8	77.3	13.2	25.7	72.7	0.0	1.5	0.0	0.28	1.80

Site number	Period of record	Latitude (decimal degrees)	Longitude (decimal degrees)	Drainage area (square miles)	Amount of upstream reservoir storage (acre-feet)	Mean basin elevation (feet)	Mean basin average annual precipitation (inches)	Area covered by developed land (percent)	Area covered by barren land (percent)
09222500	1896–97	41.583289	-110.000703	2,086.12	46,470	7,510	14.6	1.1	2.4
09223000	1952 to current year	42 110494	-110.709621	128.62	0	8,476	28.3	0.0	0.0
09223385	2007 to current year	41 963367	-110.660106	239.92	42,393	8,177	24.9	0.2	0.0
09223500	1945–75	41.857168	-110 563226	301.85	42,393	8,049	23.4	0.3	0.0
09224000	1917–33, 1945–49	41.783281	-110 534058	390.05	42,393	7,903	21.2	0.7	0.0
09224500	1896–1900	41 583289	-109 967369	2,725.16	88,863	7,499	15.1	1.1	1.9
09224700	1962 to current year	41 546072	-109.693475	2,980.53	88,863	7,407	14.5	1.2	2.1
09225000	1947–62	41 379129	-109.611529	3,651.87	88,863	7,275	13.6	1.0	2.0
09226000	1942–72	41.006337	-110 270988	56.82	0	10,318	30.8	1.1	9.0
09226500	1948–70	40 944393	-110 179321	28.55	0	10,462	30.0	0.1	11.2
09227000	1948–62	40 944394	-110 161821	7.22	0	9,352	23.2	0.5	1.5
09227500	1948–62	40 947170	-110 217377	22.21	0	10,671	32.3	0.1	17.4
09228000	1942–54	41.049953	-110.050708	1.62	0	7,684	11.7	0.0	1.4
09228500	1943–83	40 946340	-110.066265	53.74	0	10,368	29.2	0.1	11.3
09229500	1928–93, 2001 to current year	41.012459	-109.672925	527.50	0	8,525	18.9	0.8	4.8
09230300	2007 to current year	40 990278	-109.654528	22.76	0	7,181	13.2	3.4	0.4
09232000	1942–61	40.886065	-109 903485	42.45	0	10,065	27.6	0.4	6.1
09232500	1946–61	40 933292	-109.656257	99.34	13,700	8,954	21.8	0.6	3.0
09233000	1949–54	40.838844	-109.831262	12.21	0	10,274	29.1	0.0	12.2
09233500	1946–49	40.845789	-109.802650	9.25	0	10,160	29.0	0.0	8.4
09234000	1946–55	40.893014	-109.592365	111.19	0	8,993	22.5	0.7	2.1
09234500	1950–	40 908293	-109.422914	15,109.92	562,684	7,431	13.9	0.8	1.6
09234700	1971–76	40 969682	-109.237352	140.65	0	7,354	13.9	0.4	0.0
09235000	1912–15	40.862182	-109 139570	15,412.28	4,349,684	7,425	13.9	0.8	1.6
09235100	1986–90	40.795238	-109.091235	111.27	0	7,719	17.6	0.5	0.0
09235300	1975–81	41.014962	-108.644831	195.94	0	7,364	12.5	0.4	0.1
09235450	1977–81	40.761908	-108.726499	822.08	0	7,163	12.4	0.2	0.8
09235490	1995	40.722186	-108.757889	924.71	0	7,087	12.3	0.3	0.8
09235600	1957–93	40.768015	-109 319022	24.72	0	8,142	20.3	1.1	0.0
09235800	1958–82	40.673573	-109.051511	106.92	0	7,723	17.3	0.5	0.0
09236000	1952–65, 1966–86	40.043873	-107.072273	22.22	6,088	10,892	45.8	0.0	8.3
09236500	1939–44	40.070818	-106 997827	42.86	15,168	10,627	43.5	0.0	5.5
09237450	1998 to current year	40 269148	-106.880881	206.72	22,309	8,994	28.5	0.4	1.3
09237500	1939–44, 1956–72, 1984 to current year	40 285259	-106.831435	227.51	22,309	8,893	27.9	0.4	1.2
09237800	1965–73	40 295815	-106.801435	40.60	0	9,285	36.3	0.0	0.1
09238000	1952–57	40 243870	-107.015328	13.45	0	8,750	26.3	0 2	0.3
09238300	1972–75	40 395536	-106.649764	0.75	0	9,763	47.0	0.0	0.4
09238350	1972–75	40.429424	-106.676710	6.14	0	10,071	52.2	0.0	0.0
09238500	1920–22, 1965–73, 1978–87	40.408035	-106.786992	42.12	0	9,487	47.1	0.9	0.0
09238700	1984–86	40.473312	-106.680044	0.73	0	10,074	52.9	0.0	0.0
09238705	1987–95	40.473590	-106.680044	0.73	0	10,074	52.9	0.0	0.0
09238710	1985–95	40.476645	-106.687544	0.96	0	10,134	53.0	0.0	0.0
09238750	1985–95	40.498312	-106.692267	1.50	0	10,220	54.4	0.0	0.0
09238770	1985–95	40.493034	-106.692545	2.99	0	10,165	54.1	0.0	0.1
09238800	1984–94	40.497200	-106.698934	0.03	0	9,863	52.4	0.0	19.4
09238900	1966–72, 1982 to current year	40.474978	-106.786993	26.10	0	9,698	50.2	0.2	0.4
09239400	1965–72	40.493310	-106.805327	6.46	0	8,715	39.9	2.0	0.1
09239500	1904–06, 1909 to current year	40.483588	-106.832271	565.60	29,731	8,780	31.5	1.2	0.5
09240500	1911–18	40.755528	-106.810046	63.56	0	9,763	46.4	0.0	7.6
09240800	1966–73	40.745251	-106.807268	34.00	0	9,712	46.0	0.0	4.5
09240900	1988–93, 1998–2003	40.743306	-106.855325	123.23	0	9,517	43.7	0.0	5.2
09241000	1910–22, 1930–91, 1998–2003	40.717473	-106 915883	216.26	28,721	9,105	38.2	0.2	3.1

Area covered by deciduous forest (percent)	Area covered by evergreen forest (percent)	Area covered by mixed forest (percent)	Area covered by shrubs young or stunted trees (percent)	Area covered by grass or herbaceous land (percent)	Sedimentary, clastic lithology, Mesozoic (percent)	Sedimentary, clastic (continental) lithology, Tertiary (percent)	Igneous or metamorphic lithologies (percent)	Sedimentary, mixed (continental and marine) lithology (percent)	Sedimentary, carbonate (marine) lithology (percent)	Population density (people per square mile)	Road density (miles per square mile)
1.8	12.1	0.8	66.1	8.4	15.1	74.5	0.0	9.9	0.5	0.54	1.61
2.1	53.2	1.1	26.6	13.9	45.6	50.2	0.0	0.1	4.1	0.08	1.31
4.2	35.5	2.8	23.6	26.9	34.6	61.6	0.0	0.1	3.8	0.10	1.47
4.1	28.6	2.4	27.0	30.1	34.3	62.3	0.0	0.1	3.4	0.12	1.43
3.7	22.5	1.9	37.6	27.2	48.0	49.0	0.0	0.0	2.9	0.65	1.66
1.9	12.5	0.9	63.8	10.7	19.5	72.1	0.0	7.6	0.8	0.61	1.67
1.8	11.4	0.8	66.2	10.0	17.9	74.5	0.0	6.9	0.7	0.57	1.68
1.4	9.3	0.7	71.2	8.8	14.8	79.0	0.0	5.6	0.6	0.51	1.59
3.2	61.6	0.2	15.9	8.0	7.5	28.1	0.0	62.8	1.6	0.12	1.08
0.2	68.9	0.0	10.9	7.7	8.2	0.0	0.0	91.2	0.6	0.12	0.20
2.2	66.4	0.2	24.3	0.8	0.0	0.0	0.0	86.9	13.1	0.12	1.00
0.2	61.2	0.0	11.1	9.5	12.3	0.6	0.0	84.7	2.4	0.12	0.15
0.0	0.0	0.0	97.8	0.8	15.1	84.9	0.0	0.0	0.0	0.19	0.04
0.1	69.2	0.0	10.9	6.9	19.9	0.2	0.0	75.6	4.2	0.12	0.21
1.0	28.9	0.1	50.9	5.2	8.0	55.9	0.0	32.5	3.6	0.17	1.26
0.6	7.9	0.0	53.2	9.2	39.7	60.3	0.0	0.0	0.0	0.19	2.57
0.3	79.3	0.2	9.1	2.6	1.0	9.0	0.0	89.1	0.9	0.17	0.44
0.8	62.4	0.3	26.5	4.7	10.0	9.3	0.0	54.0	26.8	0.18	1.02
0.0	72.2	0.0	10.3	4.3	3.8	0.3	0.0	95.8	0.0	0.19	0.00
0.1	78.1	0.0	9.7	3.1	0.0	0.0	0.0	100.0	0.0	0.19	0.00
0.8	83.1	0.2	10.8	1.2	10.1	5.6	0.0	84.3	0.0	0.19	0.77
0.9	10.2	0.2	72.4	8.0	14.2	72.6	4.6	6.9	1.6	0.54	1.52
0.5	7.8	0.0	81.0	9.9	16.0	83.1	0.5	0.3	0.0	0.07	1.21
0.9	10.5	0.2	72.2	8.0	14.1	72.2	4.6	7.4	1.6	0.53	1.51
4.3	42.2	1.0	50.2	0.2	4.6	48.3	0.0	44.3	2.8	0.27	1.49
0.1	1.3	0.0	96.2	1.6	0.0	100.0	0.0	0.0	0.0	0.04	1.27
0.2	2.5	0.0	93.9	2.0	2.6	94.9	0.0	1.5	0.9	0.04	1.23
0.2	4.3	0.0	92.3	1.9	2.4	95.1	0.0	1.7	0.8	0.04	1.23
11.6	53.3	3.3	30.1	0.0	7.4	46.1	0.0	44.8	1.6	0.30	0.84
4.5	34.0	1.0	58.0	0.2	3.3	52.9	0.0	40.0	3.8	0.28	1.69
6.7	41.7	0.2	0.0	36.9	3.4	49.8	46.8	0.0	0.0	0.62	0.18
12.3	43.7	1.8	0.0	31.5	10.3	42.5	47.2	0.0	0.0	0.56	0.55
24.7	23.5	3.4	23.7	11.5	45.2	40.5	12.5	1.8	0.0	0.50	1.09
26.0	23.3	3.5	23.4	10.6	41.6	44.2	12.7	1.6	0.0	0.52	1.23
12.9	79.4	1.2	0.4	3.4	1.1	3.1	95.9	0.0	0.0	0.49	0.22
42.2	22.7	9.9	17.9	3.7	87.6	11.8	0.6	0.0	0.0	0.54	1.49
5.5	22.4	0.0	0.0	62.8	0.0	0.0	100.0	0.0	0.0	1.70	0.65
7.6	70.9	0.0	0.0	8.3	0.0	0.1	99.9	0.0	0.0	2.65	0.34
12.6	59.4	1.0	0.6	17.4	0.1	1.7	98.2	0.0	0.0	2.27	0.44
13.7	73.1	0.0	0.0	2.0	0.0	0.0	97.9	0.0	0.0	2.72	0.24
13.7	73.1	0.0	0.0	2.0	0.0	0.0	97.9	0.0	0.0	2.72	0.24
14.9	67.5	0.0	0.0	5.0	0.0	0.2	99.1	0.0	0.0	10.96	0.41
14.2	48.5	0.0	0.0	19.2	0.0	0.0	100.0	0.0	0.0	15.75	1.75
15.4	61.9	0.0	0.0	7.7	0.0	0.0	99.7	0.0	0.0	15.67	0.37
4.1	3.1	0.0	0.0	10.2	0.0	0.0	100.0	0.0	0.0	15.75	0.00
16.4	52.1	0.5	1.4	19.5	0.0	0.3	99.6	0.0	0.0	11.40	0.58
55.4	25.0	5.3	2.2	8.6	0.0	1.5	98.5	0.0	0.0	3.43	0.93
26.3	36.7	2.8	13.5	8.0	22.7	33.2	42.4	1.7	0.0	2.35	1.36
9.1	51.6	1.0	2.8	24.5	7.3	0.0	92.6	0.1	0.0	0.35	0.68
7.3	70.7	2.5	0.9	9.0	0.9	0.8	93.1	5.1	0.0	0.35	0.08
11.0	57.8	2.2	3.2	17.1	7.0	0.3	90.5	2.2	0.0	0.35	0.53
19.3	49.5	4.8	6.5	12.5	10.5	18.9	68.9	1.7	0.0	0.38	0.95

Site number	Period of record	Latitude (decimal degrees)	Longitude (decimal degrees)	Drainage area (square miles)	Amount of upstream reservoir storage (acre-feet)	Mean basin elevation (feet)	Mean basin average annual precipitation (inches)	Area covered by developed land (percent)	Area covered by barren land (percent)
09242500	1904–27, 1990 to current year	40.514698	-106 953941	448.30	28,721	8,663	35.8	0.3	1.8
09243700	1976–81, 1982–2001	40 385533	-106 993107	23.61	0	7,661	21.3	1.0	0.2
09243800	1976–81, 1982–83, 1985–2001	40 345811	-107.085053	8.64	0	7,382	21.4	0.1	0.1
09243900	1976–81, 1982–2001	40 390255	-106 994774	17.44	0	7,203	20.8	0.9	0.3
09244100	1955–73	40 334145	-107 139221	34.55	0	8,291	26.0	0.4	0.1
09244300	1958–66	40.446920	-107 145611	26.16	0	7,146	21.3	0.7	0.1
09244400	1965–72	40.489141	-107 159778	1,391.66	58,452	8,423	31.1	0.9	0.9
09244410	1965–86	40.488308	-107 159778	1,391.65	58,452	8,423	31.1	0.9	0.9
09244415	1980–83	40 383588	-107 193389	4.20	0	7,751	24.6	0.0	0.1
09244460	1977–81	40 382476	-107 280891	4.86	0	7,360	24.0	0.0	0.1
09244464	1977–81	40 391087	-107 271446	8.08	0	7,568	24.7	0.0	0.2
09244470	1976–81	40.468308	-107 247002	13.65	0	6,782	19.5	0.0	0.0
09244490	2004 to current year	40 518028	-107 399784	1,574.36	58,452	8,248	29.9	1.0	0.8
09244500	1942–44, 1958–73	40.732194	-107 169501	44.30	0	8,605	31.7	0.0	0.6
09245000	1953–96	40.669694	-107 285059	67.77	0	8,412	30.9	0.0	0.4
09245500	1910, 1920, 1958–73	40.680527	-107 287281	22.03	0	8,424	31.6	0.0	0.3
09246200	1995 to current year	40 591639	-107 320893	171.49	0	7,944	28.5	0.0	0.2
09246500	1906, 1909–18, 2008 to current year	40 531083	-107.436174	226.74	11,500	7,635	26.1	0.1	0.2
09246900	1955–60	40.756358	-107 546455	35.21	0	7,618	24.3	1.0	0.2
09246920	1984–91, 2002–05	40.743858	-107 540900	40.20	0	7,537	23.5	1.1	0.2
09247000	1903–06, 1909–18, 1943–47	40 514138	-107.541454	258.10	0	7,120	20.8	1.0	0.0
09247600	1984 to current year	40.480805	-107.614234	2,126.63	69,952	7,992	28.0	1.1	0.6
09248500	1943–47	40 222479	-107.266722	95.97	0	9,594	36.4	0.6	2.0
09248600	1956–72	40 261090	-107 295056	106.83	0	9,462	35.5	0.7	1.8
09249000	1953–71	40 312477	-107 320058	143.94	0	9,063	33.3	0.6	1.4
09249200	1965–79	40 212199	-107.442838	47.53	0	9,088	32.2	0.0	0.2
09249450	1985–86	40 295531	-107 512840	5.26	0	7,776	24.7	0.0	0.1
09249455	1985–86	40 296920	-107 520340	3.58	0	7,907	25.4	0.0	0.0
09249500	1904–06, 1909–27	40 369973	-107.609233	379.99	0	8,335	28.6	0.5	0.9
09249700	1965–67	40 218587	-107 581453	13.46	0	8,780	28.9	0.0	0.0
09249750	1984–2001	40.437194	-107.647846	459.48	0	8,195	27.7	0.6	0.8
09250000	1952–86	40 193585	-107.732291	63.68	0	7,905	24.2	0.6	0.4
09250400	1975–78	40 290251	-107.790071	40.06	0	7,556	22.2	2.1	0.1
09250507	1980–92	40 314695	-107.800072	20.09	0	7,636	22.5	2.1	0.1
09250510	1975–92	40 313306	-107.799794	7.19	0	7,295	20.7	0.1	0.0
09250600	1974–80	40 315528	-107.797850	27.48	0	7,538	22.0	1.7	0.1
09250610	1975–81	40 312472	-107.822295	7.64	0	7,261	20.6	0.0	0.1
09250700	1980–81	40 335805	-107.885631	26.52	0	7,410	21.3	0.0	0.1
09251000	1904–05, 1910–12, 1916 to current year	40 502747	-108.029804	3,381.29	69,952	7,756	25.7	1.0	0.6
09251100	1996–2003	40.460801	-108.425653	3,805.09	69,952	7,607	24.5	0.9	0.5
09251500	1912–22	40 990522	-107.044219	115.45	0	8,707	33.9	0.0	0.1
09251800	1956–65	41.042500	-106 957222	13.04	0	9,452	47.5	0.0	0.0
09251900	1956–63	41.015244	-107.022829	28.66	0	9,003	41.9	0.0	0.0
09252500	1912–20	40 976356	-107.050330	42.75	0	8,389	28.4	0.0	0.1
09253000	1943–47, 1950–99, 2001 current year	40 999410	-107 143388	250.63	0	8,555	33.2	0.0	0.1
09253400	1956–63, 1985–88	41 132222	-107.069167	13.65	0	9,547	48.7	1.7	0.8
09254500	1911–20, 1922	40.889412	-107 330616	91.45	0	8,763	33.3	0.0	0.5
09255000	1910–12, 1931 to current year	40 982466	-107.382839	151.32	0	8,376	30.4	0.1	0.3
09255400	1956–58, 1985–88	41 272184	-107.158943	5.62	0	8,922	37.2	0.0	0.0
09255500	1940–44, 1952–71	41.218018	-107 372283	200.98	0	7,786	25.9	0.3	0.0
09255900	1956–58, 1985–88	41 199963	-107 175611	9.92	0	9,278	41.0	0.0	0.1
09256000	1941–46, 1947–72, 1985–92	41.097778	-107 381944	309.43	0	7,861	26.9	0.3	0.1

Area covered by deciduous forest (percent)	Area covered by evergreen forest (percent)	Area covered by mixed forest (percent)	Area covered by shrubs young or stunted trees (percent)	Area covered by grass or herbaceous land (percent)	Sedimentary, clastic lithology, Mesozoic (percent)	Sedimentary, clastic (continental) lithology, Tertiary (percent)	Igneous or metamorphic lithologies (percent)	Sedimentary, mixed (continental and marine) lithology (percent)	Sedimentary, carbonate (marine) lithology (percent)	Population density (people per square mile)	Road density (miles per square mile)
28.3	35.6	4.1	13.2	9.2	22.9	19.2	55.9	2.0	0.0	0.49	0.90
58.0	5.0	2.0	29.7	0.1	100.0	0.0	0.0	0.0	0.0	1.34	1.16
48.1	6.3	0.0	42.1	0.2	100.0	0.0	0.0	0.0	0.0	1.39	0.56
32.9	5.3	0.0	54.3	0.1	100.0	0.0	0.0	0.0	0.0	1.19	1.28
47.4	16.1	7.5	18.5	4.1	77.2	14.3	8.4	0.0	0.0	0.39	0.93
41.2	3.8	0.0	45.3	0.2	100.0	0.0	0.0	0.0	0.0	0.08	1.71
30.5	29.1	2.9	19.0	6.9	37.1	23.6	37.8	1.4	0.0	1.80	1.26
30.5	29.1	2.9	19.0	6.9	37.1	23.6	37.8	1.4	0.0	1.80	1.26
82.3	1.2	0.0	13.6	0.3	100.0	0.0	0.0	0.0	0.0	0.08	1.64
68.3	0.6	0.0	29.2	0.0	100.0	0.0	0.0	0.0	0.0	0.08	1.13
60.7	1.2	0.0	35.8	0.1	100.0	0.0	0.0	0.0	0.0	0.08	1.69
18.1	1.8	0.0	71.7	0.2	100.0	0.0	0.0	0.0	0.0	0.87	1.25
29.7	26.0	2.6	23.7	6.1	43.4	21.8	33.5	1.3	0.0	1.77	1.29
42.0	21.6	5.9	23.3	5.3	72.9	24.9	2.2	0.0	0.0	0.12	0.55
48.5	16.0	4.7	24.3	4.3	76.9	21.7	1.5	0.0	0.0	0.12	0.51
59.0	11.4	3.2	12.9	10.1	17.6	81.8	0.6	0.0	0.0	0.11	0.22
43.0	10.4	2.8	36.4	3.8	78.7	20.5	0.8	0.0	0.0	0.12	0.80
33.5	8.5	2.1	47.1	2.9	72.7	26.7	0.6	0.0	0.0	0.17	1.06
30.9	10.8	1.9	49.7	1.1	0.0	99.7	0.3	0.0	0.0	0.08	0.84
27.9	9.9	1.7	54.1	1.0	0.0	99.7	0.3	0.0	0.0	0.08	0.85
16.9	4.2	0.6	69.9	0.8	2.8	96.8	0.4	0.0	0.0	1.93	1.46
27.9	20.8	2.2	33.1	5.0	42.7	31.4	25.0	0.9	0.0	2.08	1.35
35.4	40.3	3.4	2.8	13.1	59.6	26.2	14.2	0.0	0.0	0.18	0.33
39.6	36.5	3.5	3.4	12.2	58.3	28.2	13.5	0.0	0.0	0.17	0.34
46.1	27.8	3.4	7.9	9.6	65.0	23.5	11.5	0.0	0.0	0.15	0.45
50.0	24.9	4.7	6.4	12.7	79.4	17.4	3.3	0.0	0.0	0.08	0.08
83.9	0.9	0.0	13.8	0.1	97.0	0.0	3.0	0.0	0.0	0.15	1.10
95.1	0.5	0.8	2.9	0.0	100.0	0.0	0.0	0.0	0.0	0.15	0.47
50.1	17.1	2.8	18.3	6.1	80.9	12.8	6.3	0.0	0.0	0.13	0.62
68.2	14.8	4.4	3.1	8.4	66.6	33.4	0.0	0.0	0.0	0.08	0.62
49.7	15.5	2.5	20.9	5.4	83.2	11.6	5.2	0.0	0.0	0.14	0.71
48.8	12.4	3.5	22.0	3.4	87.6	12.2	0.2	0.0	0.0	0.08	0.96
44.4	8.2	0.0	38.7	1.0	100.0	0.0	0.0	0.0	0.0	0.09	1.39
47.9	8.6	0.0	36.3	1.6	100.0	0.0	0.0	0.0	0.0	0.08	2.08
23.5	4.9	0.0	70.5	0.0	100.0	0.0	0.0	0.0	0.0	0.04	2.71
41.2	7.6	0.0	45.6	1.2	100.0	0.0	0.0	0.0	0.0	0.07	2.29
19.2	9.7	0.0	70.2	0.3	100.0	0.0	0.0	0.0	0.0	0.04	2.09
31.2	5.3	0.0	60.6	0.8	100.0	0.0	0.0	0.0	0.0	0.04	2.00
27.2	16.7	1.8	41.2	4.0	52.6	30.2	16.4	0.6	0.2	1.38	1.33
24.3	15.8	1.6	46.3	3.6	47.8	36.4	14.6	0.7	0.5	1.23	1.37
32.6	44.0	7.4	6.6	6.9	11.5	31.7	56.1	0.6	0.0	0.13	0.89
5.0	74.9	2.0	0.2	17.3	0.0	0.0	100.0	0.0	0.0	0.12	0.58
19.1	60.1	7.6	1.8	11.0	0.0	1.9	97.9	0.1	0.0	0.12	0.56
41.0	23.0	6.4	23.8	3.1	27.8	61.2	11.0	0.0	0.0	0.12	0.56
33.7	37.3	6.7	12.7	6.1	11.5	38.1	49.4	1.0	0.0	0.12	0.80
2.9	71.8	0.0	1.5	20.7	0.0	0.0	78.0	21.9	0.0	0.12	1.14
35.8	32.9	8.6	14.6	6.4	17.5	78.5	4.1	0.0	0.0	0.11	0.86
37.1	25.8	6.6	24.1	4.1	15.1	79.8	5.1	0.0	0.0	0.10	0.99
1.9	86.1	0.1	6.5	5.5	0.0	0.0	65.4	34.6	0.0	0.12	1.42
5.2	13.8	0.0	74.5	3.0	9.8	77.0	11.5	1.6	0.0	0.14	1.90
3.3	90.6	0.0	0.7	5.2	0.0	0.0	89.9	10.1	0.0	0.12	0.92
11.8	20.8	0.9	59.7	3.3	14.6	68.5	15.5	1.4	0.1	0.14	1.64

Site number	Period of record	Latitude (decimal degrees)	Longitude (decimal degrees)	Drainage area (square miles)	Amount of upstream reservoir storage (acre-feet)	Mean basin elevation (feet)	Mean basin average annual precipitation (inches)	Area covered by developed land (percent)	Area covered by barren land (percent)
09256500	1914–16, 1918–22	41.024687	-107.444507	352.18	0	7,784	26.1	0.4	0.0
09257000	1910–23, 1938–98	41.028298	-107.549233	1,069.99	0	7,996	27.8	0.3	0.1
09258000	1953–93	40.915522	-107.521732	24.62	0	8,066	28.5	0.0	0.3
09258980	2004 to current year	41.068159	-107.631596	1,008.89	0	6,961	13.1	0.3	0.6
09259000	1915–16, 1918, 1987–91	41.046909	-107.650903	1,013.48	0	6,959	13.1	0.3	0.6
09259700	1961–68	41.003020	-107.920356	3,011.69	0	7,246	18.0	0.3	0.6
09259950	1950–69	40.607467	-108.337039	3,688.10	0	7,116	17.1	0.3	0.6
09259990	1987–91	40.619967	-108.368985	239.29	0	6,548	12.8	0.0	0.0
09260000	1904, 1921 to current year	40.547189	-108.424265	4,040.79	0	7,060	16.7	0.3	0.6
09260050	1982–94, 1996 to current year	40.451634	-108.525101	7,936.43	69,952	7,313	20.4	0.6	0.6
09260500	1950–56, 1960–61	40.559130	-109.054843	121.60	0	7,577	16.3	0.7	0.7
09261000	1903–06, 1914–15, 1946 to current year	40.409408	-109.235404	25,504.79	4,419,636	7,354	15.9	0.7	1.2
09261500	1946–55	40.704124	-109.596531	23.48	6,249	9,491	27.7	1.2	0.0
09261700	1979 to current year	40.588848	-109.465414	77.28	6,249	8,318	22.0	2.1	1.3
09262500	1949–55	40.758291	-109.534029	9.24	0	9,287	26.7	0.5	0.0
09263000	1946–52	40.716070	-109.505694	29.05	0	9,251	27.0	0.8	0.0
09263500	1940–65	40.408573	-109.340963	259.01	32,249	7,276	17.4	1.5	2.0
09264000	1944–54	40.733290	-109.678479	26.72	0	9,932	28.9	2.1	0.9
09264500	1944–55	40.733290	-109.703480	20.39	0	10,458	31.8	0.7	10.3
09265000	1946–69	40.743290	-109.622366	11.73	6,249	9,637	28.0	0.6	0.0
09265300	1965–75	40.679679	-109.660978	56.11	0	10,015	29.6	1.4	4.2
09265500	1941–45	40.588847	-109.622919	100.34	0	9,477	27.3	1.1	2.4
09266000	1944–45, 1954–55	40.586069	-109.622919	100.42	0	9,475	27.3	1.1	2.4
09266500	1911–12, 1914–17, 1919 to current year	40.577458	-109.622086	101.44	0	9,453	27.2	1.1	2.5
09267100	1969–72	40.537736	-109.609863	110.26	0	9,256	26.2	1.1	2.4
09268000	1940–75	40.626345	-109.820149	44.56	0	10,349	31.7	0.2	8.4
09268500	1947–89	40.642733	-109.810982	8.81	0	10,126	30.8	0.0	3.5
09268900	1961–89	40.659401	-109.750981	7.50	0	10,024	30.4	1.8	0.8
09269000	1947–63	40.649956	-109.761814	10.33	0	9,831	29.5	1.5	0.6
09269500	1950–52	40.633290	-109.764592	18.25	0	9,605	28.3	0.9	0.4
09270000	1904, 1941–45, 1954–69	40.569402	-109.697644	97.32	0	9,700	28.6	0.6	4.2
09270500	1954–89	40.526348	-109.605696	115.82	0	9,279	26.6	0.8	3.7
09271000	1900–04, 1939–65	40.517181	-109.596529	238.52	0	9,141	25.8	1.0	3.0
09271400	2000–03	40.433572	-109.466246	335.47	38,170	8,189	21.3	2.9	3.3
09271450	2000–03	40.398017	-109.429577	368.70	38,170	7,930	20.1	3.3	3.2
09271500	1947–83	40.374684	-109.408188	379.62	38,170	7,851	19.8	3.4	3.2
09271550	1991–2002	40.358018	-109.387631	380.79	38,170	7,841	19.7	3.4	3.2
09272400	2009 to current year	40.084686	-109.676805	26,801.50	4,501,905	7,312	15.8	0.7	1.4
09273000	1930–33, 1935–54	40.624945	-110.889613	37.38	0	10,179	39.9	0.7	13.2
09273200	1965–68	40.622168	-110.892390	41.32	0	10,114	39.6	0.7	12.4
09273500	1950–68	40.536055	-110.867387	7.96	0	9,955	35.8	0.0	15.3
09274000	1922–23, 1929–30, 1946–63	40.533278	-110.867387	79.92	0	9,835	37.1	0.5	9.8
09274900	1989–94	40.450223	-111.004332	37.05	0	9,216	32.1	0.6	0.1
09275000	1950–68, 1974–81	40.448556	-110.975720	43.81	0	9,149	31.6	0.5	0.1
09275500	1945–94	40.450223	-110.884330	61.56	0	8,860	29.3	0.4	0.1
09276000	1946–84	40.471056	-110.918775	10.70	0	9,159	31.7	2.7	0.2
09276500	1922–23	40.452723	-110.882664	19.47	0	9,119	30.8	2.2	0.3
09276600	1990–2003	40.461612	-110.836830	89.46	0	8,844	28.9	0.9	0.2
09277000	1953–61	40.415225	-110.782662	236.54	0	9,101	30.6	0.6	3.9
09277500	1918 to current year	40.300230	-110.602381	356.67	0	8,707	26.9	0.8	3.0
09277501	1919–67	40.300230	-110.602381	356.67	0	8,707	26.9	0.8	3.0
09277800	1965–84, 1988–94	40.557445	-110.697941	99.07	0	10,368	37.4	0.0	16.1

Area covered by deciduous forest (percent)	Area covered by evergreen forest (percent)	Area covered by mixed forest (percent)	Area covered by shrubs young or stunted trees (percent)	Area covered by grass or herbaceous land (percent)	Sedimentary, clastic lithology, Mesozoic (percent)	Sedimentary, clastic (continental) lithology, Tertiary (percent)	Igneous or metamorphic lithologies (percent)	Sedimentary, mixed (continental and marine) lithology (percent)	Sedimentary, carbonate (marine) lithology (percent)	Population density (people per square mile)	Road density (miles per square mile)
12.1	18.7	0.8	60.4	3.1	13.8	71.3	13.6	1.2	0.1	0.14	1.68
22.4	23.6	3.2	42.0	4.0	17.8	60.1	20.6	1.4	0.0	0.12	1.23
29.9	26.3	4.2	36.8	0.6	0.0	99.1	0.9	0.0	0.0	0.08	1.65
0.1	0.9	0.0	96.5	1.1	30.3	69.0	0.0	0.0	0.1	0.12	1.37
0.1	0.9	0.0	96.4	1.1	30.1	69.1	0.0	0.0	0.1	0.12	1.38
8.2	8.9	1.2	76.1	2.0	16.6	75.0	7.3	0.8	0.0	0.11	1.21
6.7	7.5	1.0	79.8	1.9	14.3	78.8	6.0	0.7	0.0	0.09	1.23
0.0	2.0	0.0	95.9	2.0	1.8	98.2	0.0	0.0	0.0	0.04	1.01
6.1	7.4	0.9	80.8	1.8	13.1	80.1	5.5	1.0	0.1	0.09	1.22
14.8	11.8	1.2	64.0	2.7	30.1	58.7	9.8	0.9	0.5	0.64	1.29
2.3	20.0	0.3	74.3	0.1	2.9	56.3	0.0	5.6	35.2	0.36	2.16
5.2	11.8	0.5	69.9	5.8	18.6	66.7	5.8	6.1	2.8	0.53	1.42
0.6	84.2	0.2	4.7	4.2	3.4	50.7	0.0	44.9	1.0	0.35	1.36
9.8	52.5	1.1	29.6	1.4	7.7	39.7	0.0	14.4	38.2	0.38	1.75
0.1	70.3	0.3	13.8	7.5	4.4	37.2	0.0	58.4	0.0	0.35	1.25
1.1	79.6	0.7	9.1	5.7	1.4	40.0	0.0	54.3	4.2	0.35	1.34
5.9	38.6	0.6	47.0	1.1	42.3	27.8	0.0	10.4	19.4	0.44	1.78
0.0	70.8	0.0	10.9	9.4	0.0	44.8	0.0	55.2	0.0	0.35	1.84
0.0	65.1	0.0	12.1	8.4	0.0	0.5	0.0	99.5	0.0	0.35	0.24
0.0	77.7	0.1	5.7	7.0	6.8	56.5	0.0	36.7	0.0	0.35	1.43
0.1	72.5	0.2	9.9	7.7	0.0	30.7	0.0	69.3	0.0	0.35	1.03
4.1	74.0	2.1	9.4	4.5	0.4	41.7	0.0	47.4	10.6	0.35	0.87
4.1	74.0	2.1	9.4	4.5	0.4	41.7	0.0	47.3	10.6	0.35	0.87
4.0	73.8	2.0	9.7	4.4	0.4	41.4	0.0	46.8	11.4	0.35	0.86
4.1	72.2	1.9	11.6	4.1	4.1	39.5	0.0	43.1	13.3	0.35	0.88
0.3	65.6	0.2	12.1	11.3	0.0	8.0	0.0	89.2	2.8	0.35	0.38
0.7	74.0	0.5	8.8	12.3	0.0	6.4	0.0	93.6	0.0	0.35	0.44
0.1	75.0	0.1	13.7	6.2	0.0	44.5	0.0	55.5	0.0	0.35	1.81
1.2	78.9	0.9	10.4	4.8	0.0	42.4	0.0	53.3	4.3	0.35	1.67
2.8	82.0	2.0	8.1	2.9	0.7	37.6	0.0	48.5	13.2	0.35	1.25
5.3	67.2	1.8	12.6	7.0	1.2	19.1	0.0	61.1	18.6	0.35	0.76
5.3	64.7	1.6	15.8	5.9	12.4	19.1	0.0	51.3	17.1	0.35	1.03
4.6	66.7	1.6	16.0	4.7	10.2	28.4	0.0	44.9	16.6	0.35	1.00
3.3	52.9	1.2	24.6	3.4	31.9	24.4	0.0	31.9	11.8	5.66	1.67
3.0	48.3	1.1	26.5	3.1	37.3	22.9	0.0	29.0	10.7	6.18	1.83
2.9	46.9	1.0	26.9	3.0	38.8	22.6	0.0	28.2	10.4	6.10	1.86
2.9	46.8	1.0	26.9	3.0	38.8	22.7	0.0	28.1	10.4	6.08	1.86
5.1	12.4	0.5	69.2	5.6	19.5	65.6	5.5	6.3	3.0	0.61	1.44
0.5	73.4	0.5	5.5	4.4	0.0	0.0	0.0	100.0	0.0	0.08	0.93
0.5	73.3	0.6	6.8	4.1	0.0	0.0	0.0	100.0	0.0	0.07	1.13
3.5	66.7	1.9	8.8	3.5	8.2	0.8	0.0	84.7	6.2	0.08	0.54
3.4	71.4	1.4	9.5	2.9	1.7	1.9	0.0	88.8	7.5	0.07	1.09
42.4	32.8	5.9	17.9	0.2	42.9	57.1	0.0	0.0	0.0	0.01	1.50
44.6	31.8	5.7	17.1	0.2	48.6	51.4	0.0	0.0	0.0	0.01	1.39
43.5	28.1	4.2	23.4	0.2	62.1	37.9	0.0	0.0	0.0	0.01	1.30
35.2	42.2	5.1	14.2	0.1	1.0	42.5	0.0	19.5	37.0	0.02	3.22
31.7	42.2	4.3	18.7	0.4	3.7	26.2	0.0	17.5	52.6	0.04	2.62
39.9	30.5	3.9	24.3	0.3	47.9	31.8	0.0	4.0	16.4	0.02	1.50
22.8	45.1	2.3	21.6	2.5	29.9	14.8	0.0	33.9	21.4	0.05	1.25
20.5	44.9	1.9	24.1	2.0	32.6	25.8	0.0	24.2	17.4	0.06	1.26
20.5	44.9	1.9	24.1	2.0	32.6	25.8	0.0	24.2	17.4	0.06	1.26
0.3	67.1	0.6	8.0	5.3	0.1	0.0	0.0	99.9	0.0	0.08	0.01

Site number	Period of record	Latitude (decimal degrees)	Longitude (decimal degrees)	Drainage area (square miles)	Amount of upstream reservoir storage (acre-feet)	Mean basin elevation (feet)	Mean basin average annual precipitation (inches)	Area covered by developed land (percent)	Area covered by barren land (percent)
09278000	1953–92	40.548278	-110.694330	14.78	0	10,150	34.7	0.1	15.7
09278500	1950–69, 1974–88	40 545501	-110.656274	121.10	0	10,283	36.6	0.1	15.3
09278700	1974–81	40 532446	-110.623496	136.61	0	10,200	35.7	0.1	14.0
09279000	1937 to current year	40.493280	-110 578217	146.33	0	10,101	34.9	0.1	13.1
09279100	1963–94	40 311064	-110.494047	235.90	0	9,340	29.1	0.2	8.3
09279150	1970–2003	40 270510	-110.442657	623.38	0	8,847	27.0	0.6	4.9
09279500	1918–70	40 164125	-110 393488	664.33	0	8,690	26.0	0.7	4.7
09280000	1950–60	40 333288	-111 234061	11.18	0	9,190	32.8	0.3	0.0
09280400	1964–84	40 298289	-111 265173	2.98	0	9,016	33.8	0.9	0.0
09281500	1949–60	40 299956	-111 250728	3.48	0	8,919	33.6	1.1	0.0
09285000	1942–56, 1963–94	40 133288	-111.024887	213.23	1,106,500	8,219	25.4	1.2	0.1
09285500	1943–47	40 118010	-111.010442	43.92	0	8,758	26.6	0.3	0.5
09285700	1964–81	40 116900	-110.808214	363.01	1,106,500	8,305	24.6	1.0	1.4
09285900	1989–94, 2005 to current year	40 127178	-110.741824	372.23	1,106,500	8,277	24.4	1.0	1.5
09286100	1986–98	40 329950	-110.862663	31.38	0	8,660	26.8	0.2	0.1
09286500	1918–22, 1956–61	40 201343	-110.784048	87.38	5,700	8,064	22.2	0.8	0.1
09286700	1983–94	40 330784	-111.049611	47.09	15,670	8,863	28.9	0.5	0.2
09287000	1946–68, 1974–83	40 324117	-111.046000	49.56	15,670	8,831	28.7	0.5	0.2
09287500	1946–84	40 241619	-110 980720	13.84	0	8,481	25.4	0.1	0.0
09288000	1934–2003, 2005 to current year	40 200231	-110 907662	140.41	15,670	8,335	24.3	0.7	0.1
09288100	1964–81	40 146344	-110.753214	297.00	21,370	7,936	21.4	1.0	0.4
09288150	1964–86	39 993014	-110.814880	56.16	0	8,224	22.1	0.7	5.8
09288180	1968 to current year	40 154679	-110.554879	917.40	1,127,870	8,002	21.9	0.9	4.5
09288400	1989–94	40 173846	-110.429600	1,060.10	1,295,170	7,815	20.7	1.0	4.1
09288500	1908–10, 1915–68	40 161069	-110.411822	1,066.57	1,295,170	7,804	20.6	1.0	4.1
09288900	1964–86	39 989406	-110.459876	40.59	0	8,129	20.2	0.3	19.3
09289000	1918–21	40 136071	-110 209042	197.51	0	7,301	16.0	0.5	8.7
09289500	1933–34, 1942–55, 1963 to current year	40.606613	-110 527107	77.81	0	10,787	35.3	0.0	23.5
09290000	1933–34, 1943–55	40 574946	-110 517384	13.80	0	10,169	32.8	0.0	6.9
09291000	1921–34, 1942 to current year	40 556336	-110.484606	112.95	49,500	10,498	33.1	0.1	17.2
09291200	1977–84	40 501337	-110.405438	130.04	49,500	10,309	31.8	0.1	15.0
09291500	1950–55	40 597169	-110 347938	98.61	0	10,825	32.3	0.0	23.3
09292000	1996 to current year	40 546337	-110 333771	107.99	0	10,632	31.4	0.1	21.3
09292500	1944 to current year	40 511893	-110.341549	125.74	0	10,452	30.4	0.2	18.3
09293000	1943–44, 1976–81	40.444117	-110.364048	132.31	0	10,302	29.6	0.2	17.4
09293500	1904, 1907–10, 1917–20, 1976–81	40.436895	-110.364326	298.68	49,500	10,042	29.2	0.3	14.3
09293600	1976–81	40 357731	-110 313491	322.90	49,500	9,829	28.1	0.4	13.2
09293700	1976–79	40 310511	-110 294045	95.41	0	7,105	13.4	2.1	0.1
09294000	1943–55, 1976–81	40 258291	-110 222376	434.40	49,500	9,089	24.1	0.8	9.8
09294500	1900–03, 1907–36, 1976–81	40 208848	-110 117096	488.25	55,300	8,696	22.4	1.0	8.9
09295000	1899 to current year	40 200238	-110.063761	2,651.28	1,350,470	8,047	21.3	1.1	5.9
09295100	1998 to current year	40 206628	-109.859867	2,705.10	1,350,470	7,988	21.1	1.1	5.8
09295500	1951–55	40.786056	-110 239601	27.62	0	11,463	32.1	0.0	31.4
09296000	1946–55	40.630505	-110 159045	133.74	0	10,915	30.4	0.0	24.2
09296500	1950–55	40.624950	-110 131267	9.30	0	10,469	31.0	0.0	3.6
09296800	1990 to current year	40 591340	-110 114323	156.06	0	10,697	29.8	0.1	20.9
09297000	1922–27, 1930–83	40 535508	-110.063488	163.60	0	10,552	29.2	0.2	20.0
09297500	1899–1903, 1907–10, 1917–20	40 520787	-110.040988	198.77	0	10,342	28.8	0.2	16.7
09297600	1976–81	40.460789	-109.950707	216.80	0	10,083	27.7	0.4	15.3
09297700	1977–81	40.471900	-109 955707	215.76	0	10,102	27.8	0.4	15.4
09297800	1976–81	40.454956	-109 936539	249.64	0	9,834	26.9	0.5	13.3
09297900	1976–82	40.403569	-109.875981	383.94	0	9,814	27.3	0.6	12.0

Area covered by deciduous forest (percent)	Area covered by evergreen forest (percent)	Area covered by mixed forest (percent)	Area covered by shrubs young or stunted trees (percent)	Area covered by grass or herbaceous land (percent)	Sedimentary, clastic lithology, Mesozoic (percent)	Sedimentary, clastic (continental) lithology, Tertiary (percent)	Igneous or metamorphic lithologies (percent)	Sedimentary, mixed (continental and marine) lithology (percent)	Sedimentary, carbonate (marine) lithology (percent)	Population density (people per square mile)	Road density (miles per square mile)
1.4	60.5	1.2	9.4	11.4	1.8	0.0	0.0	67.2	31.0	0.08	0.81
0.7	66.8	0.9	8.3	5.8	0.6	0.5	0.0	93.8	5.1	0.08	0.15
1.2	67.7	1.2	8.4	5.5	0.6	2.4	0.0	89.2	7.7	0.08	0.19
1.9	67.2	1.3	9.0	5.3	0.6	3.9	0.0	83.7	11.8	0.08	0.23
8.2	59.8	1.4	16.9	3.8	10.4	23.4	0.0	55.1	11.0	0.10	0.42
14.8	51.2	1.6	21.6	2.6	22.6	28.6	0.0	34.7	14.1	0.08	0.93
13.9	49.4	1.5	24.0	2.4	22.1	32.1	0.0	32.6	13.3	0.16	1.02
57.9	21.6	6.8	13.2	0.1	36.1	44.4	0.0	0.0	19.4	0.03	2.18
47.8	24.9	16.7	9.6	0.0	0.0	100.0	0.0	0.0	0.0	0.05	2.20
50.7	22.3	16.2	9.5	0.0	0.0	99.8	0.0	0.0	0.2	0.05	2.13
44.0	9.6	1.9	31.7	0.2	19.4	75.0	0.0	0.4	5.1	0.00	1.79
49.3	20.8	2.3	26.9	0.0	0.0	100.0	0.0	0.0	0.0	0.00	1.80
36.8	22.9	2.2	29.0	0.2	11.4	85.3	0.0	0.3	3.0	0.01	1.59
35.9	24.1	2.1	28.8	0.2	11.1	85.7	0.0	0.2	2.9	0.01	1.57
55.2	16.6	2.8	24.9	0.0	74.6	25.4	0.0	0.0	0.0	0.03	1.33
32.1	16.8	1.9	44.8	0.1	32.5	67.5	0.0	0.0	0.0	0.04	2.22
65.8	13.5	3.6	15.5	0.1	73.4	26.6	0.0	0.0	0.0	0.00	1.90
64.9	13.4	3.4	16.7	0.1	74.1	25.9	0.0	0.0	0.0	0.00	1.92
55.8	23.7	1.3	18.8	0.3	0.5	99.5	0.0	0.0	0.0	0.00	1.23
53.4	14.4	1.9	28.8	0.1	30.0	70.0	0.0	0.0	0.0	0.00	1.74
35.7	18.2	1.5	40.7	0.1	29.0	71.0	0.0	0.0	0.0	0.03	2.03
9.0	38.5	1.4	44.6	0.0	0.0	100.0	0.0	0.0	0.0	0.03	1.33
26.9	28.6	1.6	33.8	0.1	13.9	84.8	0.0	0.1	1.2	0.03	1.58
23.7	28.8	1.4	37.4	0.1	14.1	84.7	0.0	0.1	1.0	0.07	1.68
23.6	28.7	1.4	37.6	0.1	14.1	84.8	0.0	0.1	1.0	0.08	1.69
1.5	43.3	0.4	34.9	0.2	0.0	100.0	0.0	0.0	0.0	0.08	0.82
1.0	39.3	0.1	49.4	0.3	0.6	99.4	0.0	0.0	0.0	0.09	1.05
0.1	51.4	0.3	12.9	10.8	0.0	0.0	0.0	100.0	0.0	0.08	0.00
0.9	81.8	0.4	3.8	3.0	0.0	4.3	0.0	89.7	6.0	0.08	0.18
0.7	60.2	0.7	10.5	8.4	0.9	3.0	0.0	94.8	1.3	0.08	0.13
2.4	60.5	0.8	11.4	7.5	1.8	8.7	0.0	82.7	6.8	0.08	0.24
0.2	48.3	0.3	14.3	12.1	0.0	2.7	0.0	97.1	0.1	0.08	0.09
2.1	49.8	0.6	13.7	11.1	0.0	5.3	0.0	91.2	3.4	0.08	0.17
2.4	53.9	0.7	13.1	10.0	0.0	15.4	0.0	80.9	3.8	0.08	0.38
2.4	52.8	0.7	15.6	9.5	0.2	16.0	0.0	80.2	3.6	0.11	0.47
3.1	56.1	0.8	16.3	7.5	0.9	17.5	0.0	77.0	4.6	0.14	0.56
3.1	53.9	0.8	18.0	7.0	0.8	18.2	0.0	76.7	4.2	0.17	0.67
1.8	17.5	0.1	49.3	0.1	9.9	61.9	0.0	28.2	0.0	0.68	2.40
2.7	44.4	0.6	26.3	5.2	3.0	30.3	0.0	63.6	3.1	0.33	1.08
2.4	39.8	0.6	30.4	4.6	8.0	32.7	0.0	56.6	2.8	0.46	1.21
13.5	36.3	1.1	35.1	1.6	13.9	63.3	0.0	18.6	4.2	0.27	1.40
13.3	35.6	1.1	35.6	1.5	13.9	63.7	0.0	18.2	4.2	0.27	1.41
0.0	28.9	0.0	17.4	21.4	0.0	0.0	0.0	100.0	0.0	0.08	0.00
0.2	44.8	0.3	13.9	15.2	0.0	3.1	0.0	96.9	0.0	0.08	0.01
0.9	58.7	0.1	23.4	12.9	0.0	78.0	0.0	22.0	0.0	0.08	0.00
1.0	48.6	0.5	13.8	13.8	0.0	12.3	0.0	87.5	0.2	0.10	0.08
1.3	49.5	0.6	13.9	13.2	0.0	12.7	0.0	87.1	0.2	0.15	0.20
3.4	52.7	0.9	13.0	11.8	0.0	21.3	0.0	77.6	1.1	0.16	0.33
3.7	53.6	0.9	13.9	10.8	0.4	25.8	0.0	72.8	1.0	0.21	0.44
3.7	53.7	0.9	13.7	10.9	0.4	25.9	0.0	72.7	1.0	0.21	0.43
5.0	54.8	1.1	13.8	9.6	1.1	29.9	0.0	65.2	3.8	0.24	0.58
4.1	55.0	0.9	14.8	9.8	0.9	24.0	0.0	71.6	3.5	0.31	0.68

Site number	Period of record	Latitude (decimal degrees)	Longitude (decimal degrees)	Drainage area (square miles)	Amount of upstream reservoir storage (acre-feet)	Mean basin elevation (feet)	Mean basin average annual precipitation (inches)	Area covered by developed land (percent)	Area covered by barren land (percent)
09298000	1950–81	40.567454	-109 961541	14.76	0	9,183	26.2	0.4	0.0
09298500	1946–55	40.636064	-109 967375	89.51	0	10,646	31.7	0.3	14.0
09299000	1946–55	40.616621	-109 939596	11.77	0	9,846	29.8	1.9	0.3
09299400	1976–81	40 586899	-109 927651	115.34	0	10,300	30.6	0.4	11.0
09299500	1899–1903, 1907–10, 1913 to current year	40 593565	-109 932374	115.18	0	10,304	30.6	0.4	11.1
09299600	1976–81	40 532455	-109 923207	120.49	0	10,183	30.0	0.5	10.6
09299700	1976–81	40.467178	-109 914039	123.79	0	10,087	29.6	0.5	10.3
09299900	1976–79	40 377736	-109.821535	73.35	0	7,229	16.3	1.2	0.2
09300000	1943–45, 1950–55	40 333293	-109.847924	83.65	0	7,012	15.4	1.6	0.2
09300500	1899–1904, 1907–10, 1917–20, 1943–58, 1976–81	40 301904	-109.853201	557.43	0	8,736	22.6	1.2	8.4
09301000	1951–58	40.463841	-110 161824	40.29	0	9,043	23.1	0.8	3.2
09301200	1976–81	40 238850	-109.852367	655.54	18,800	7,726	17.3	2.3	5.4
09301500	1899–1904, 1976–81, 1998 to current year	40 233572	-109.803755	1,072.57	29,900	7,638	17.5	2.2	4.6
09302000	1942 to current year	40 215517	-109.783476	3,777.08	1,380,370	7,879	20.0	1.4	5.5
09302400	1956–65	39 997761	-107 231165	20.31	0	10,792	44.9	0.0	5.1
09302420	1965–73	40.046925	-107 311167	62.68	0	10,370	41.8	0.0	3.1
09302450	1964–89	40.050256	-107.468948	21.66	0	8,975	29.9	0.0	0.1
09302500	1903–06, 1973–84	40.038312	-107.488115	59.85	0	9,808	37.8	0.0	2.1
09302800	1903–06, 1956–72	40.035534	-107 520894	219.76	0	9,748	37.1	0.3	1.8
09303000	1910–16, 1919–21, 1952–2001	39 987477	-107.614508	259.11	0	9,549	35.5	0.4	1.5
09303300	1975–95	39.843315	-107.334778	52.60	0	10,583	42.2	0.0	1.1
09303320	1975–89	39.842760	-107.336722	7.55	0	10,654	41.8	0.0	0.0
09303340	1976–77	39.818037	-107.391723	11.13	0	10,593	42.2	0.0	0.6
09303400	1976–95	39.864146	-107 533948	129.81	0	10,266	40.0	0.0	0.8
09303500	1903–06, 1910–15, 1942–47, 1967–92	39 921646	-107 551727	153.50	0	10,080	38.7	0.0	0.7
09304000	1919–20, 1952–97	39 974422	-107.625341	178.08	0	9,857	37.1	0.0	0.6
09304100	1955–64	39 968866	-107.646731	34.23	7,658	8,355	25.5	0.0	0.0
09304150	1970–79	39 931087	-107.770068	57.97	0	8,587	26.9	0.0	0.1
09304200	1961 to current year	40.004974	-107.825349	647.07	7,658	9,215	32.7	0.3	0.8
09304300	1957–68	40.091363	-107.770069	25.47	0	8,047	24.2	0.0	0.1
09304500	1901–06, 1909 to current year	40.033585	-107.862295	745.72	7,658	8,972	31.1	0.4	0.7
09304600	1978–85	40.033307	-107 918686	810.51	7,658	8,826	30.2	0.7	0.7
09304800	1961 to current year	40.013305	-108.093136	1,024.77	7,658	8,460	28.0	0.9	0.6
09305500	1952–57	39.732199	-107.938406	9.01	0	8,237	23.1	0.0	0.2
09306007	1974–98	39.826085	-108.183138	177.13	0	7,634	20.0	0.8	0.9
09306015	1974–76, 1977–82	39.788863	-108.173693	24.13	0	7,663	18.3	0.0	2.1
09306022	1976–85	39.813307	-108 183972	59.94	0	7,589	18.3	0.0	1.5
09306025	1974–76, 1977–82	39.783585	-108 189805	14.35	0	7,604	18.2	0.0	1.1
09306028	1974–82	39.812474	-108 183972	15.76	0	7,531	18.2	0.0	1.0
09306033	1974–76, 1977–82	39.785252	-108 209806	1.24	0	7,161	17.6	0.0	0.0
09306036	1974–86	39.824973	-108 199250	3.68	0	6,936	18.0	0.0	0.0
09306039	1974–85	39.826640	-108 207584	1.17	0	6,740	18.2	0.0	0.0
09306042	1974–84, 1985–92	39.833584	-108 220640	1.04	0	6,669	17.6	0.0	0.0
09306045	1980–82, 1985	39.835529	-108 221195	255.37	0	7,574	19.4	0.6	1.1
09306050	1974–76, 1978–82	39.793862	-108 228418	6.58	0	7,311	17.9	0.0	0.0
09306052	1974–85	39.814140	-108 243696	7.93	0	7,214	17.9	0.0	0.0
09306058	1974–85	39.837195	-108.244252	48.30	0	7,473	18.4	0.0	1.9
09306061	1974–87	39.850528	-108.258975	309.17	0	7,542	19.2	0.5	1.2
09306175	1975–83	39.871362	-108.287587	102.99	0	7,341	18.6	0.0	2.3
09306200	1964 to current year	39 921083	-108 297588	505.72	0	7,418	18.8	0.4	1.6
09306202	1977–81	39 933028	-108 317033	0.21	0	6,560	15.1	0.0	0.0
09306203	1977–81	39 936639	-108 298699	515.14	0	7,405	18.8	0.4	1.6

Area covered by deciduous forest (percent)	Area covered by evergreen forest (percent)	Area covered by mixed forest (percent)	Area covered by shrubs young or stunted trees (percent)	Area covered by grass or herbaceous land (percent)	Sedimentary, clastic lithology, Mesozoic (percent)	Sedimentary, clastic (continental) lithology, Tertiary (percent)	Igneous or metamorphic lithologies (percent)	Sedimentary, mixed (continental and marine) lithology (percent)	Sedimentary, carbonate (marine) lithology (percent)	Population density (people per square mile)	Road density (miles per square mile)
11.6	78.0	4.5	3.4	1.9	4.4	38.8	0.0	18.2	38.6	0.28	0.52
0.2	56.1	0.1	13.0	14.7	0.0	3.1	0.0	96.9	0.0	0.23	0.43
3.2	82.5	0.6	4.5	3.9	0.2	49.0	0.0	50.7	0.0	0.81	1.78
2.4	60.5	0.7	11.3	11.9	0.4	9.8	0.0	87.8	2.1	0.36	0.64
2.4	60.5	0.7	11.3	11.9	0.4	9.8	0.0	87.9	1.9	0.36	0.64
2.9	59.7	0.7	12.5	11.4	0.5	11.5	0.0	85.0	3.1	0.38	0.67
2.8	58.8	0.7	13.7	11.1	0.5	12.7	0.0	83.8	3.0	0.40	0.71
9.8	38.3	0.3	43.6	0.2	55.9	40.6	0.0	0.0	3.5	0.87	1.56
8.6	33.9	0.3	42.0	0.2	49.4	47.5	0.0	0.0	3.1	0.98	1.65
4.2	45.2	0.7	23.1	6.9	11.8	33.7	0.0	51.6	2.8	0.69	1.08
9.6	49.8	1.8	29.2	5.1	0.0	21.3	0.0	76.5	2.2	0.28	1.11
2.6	32.0	0.5	31.9	4.3	7.0	46.8	0.0	45.7	0.6	2.24	1.59
3.2	33.6	0.5	31.9	4.2	10.8	46.5	0.0	41.1	1.6	1.81	1.59
10.4	34.9	0.9	34.5	2.3	13.1	59.0	0.0	24.4	3.4	0.71	1.46
2.5	41.6	0.2	0.0	40.2	0.0	7.1	92.9	0.0	0.0	0.62	0.10
8.2	54.1	0.9	0.3	27.8	0.0	20.3	67.8	2.7	9.2	0.60	0.18
49.9	15.4	3.7	12.8	17.5	87.9	10.9	0.1	0.0	1.2	0.08	0.37
20.5	45.9	11.3	0.9	14.2	0.0	38.9	49.6	1.8	9.7	0.10	0.13
22.4	46.6	5.4	2.7	17.0	11.9	27.5	41.5	1.9	17.3	0.25	0.30
27.6	42.3	5.8	3.2	15.5	14.6	27.2	35.8	2.1	20.3	0.22	0.39
4.1	46.1	0.1	0.1	39.0	0.0	0.6	57.1	0.0	42.3	0.61	0.35
1.0	39.0	0.1	0.0	57.3	0.0	0.0	5.7	0.0	94.3	0.62	1.71
2.3	47.3	0.1	0.0	46.3	0.0	0.0	7.8	0.0	92.2	0.84	1.23
9.9	49.1	1.4	0.6	33.0	0.0	0.4	40.6	0.0	59.0	0.54	0.47
15.2	46.8	2.8	0.8	28.4	0.0	2.2	38.5	1.6	57.8	0.47	0.43
21.0	43.0	4.1	1.1	24.8	0.0	7.8	34.5	2.4	55.3	0.42	0.44
65.3	9.2	6.0	8.8	5.7	36.5	10.1	0.5	0.0	52.9	0.08	0.81
47.9	15.7	7.9	7.6	18.5	0.0	2.3	0.0	0.0	97.7	0.21	0.57
33.2	34.3	5.7	4.7	16.0	10.9	15.3	24.4	1.6	47.8	0.25	0.53
63.3	8.8	5.7	11.3	4.4	92.9	0.0	0.0	0.0	7.1	0.08	1.18
34.1	30.8	5.2	8.1	14.0	21.5	14.0	21.2	1.4	41.9	0.22	0.63
33.9	29.2	4.8	10.3	13.0	27.3	13.2	19.5	1.3	38.7	0.44	0.72
32.5	28.4	3.9	15.1	10.5	30.3	21.0	15.4	1.0	32.3	0.50	0.92
70.6	7.0	1.0	13.7	4.1	94.1	0.0	0.0	0.0	5.9	0.46	1.71
29.7	33.3	0.1	29.6	0.9	16.1	83.3	0.0	0.0	0.6	0.16	1.61
17.4	31.5	0.0	47.5	0.1	0.0	100.0	0.0	0.0	0.0	0.13	1.70
17.6	36.9	0.0	42.7	0.1	0.0	100.0	0.0	0.0	0.0	0.13	1.63
15.2	43.2	0.0	39.1	0.1	0.0	100.0	0.0	0.0	0.0	0.13	1.53
13.9	46.4	0.0	37.4	0.1	0.0	100.0	0.0	0.0	0.0	0.13	1.60
0.0	2.5	0.0	96.0	1.5	0.0	100.0	0.0	0.0	0.0	0.12	0.03
0.0	47.7	0.0	51.8	0.5	0.0	100.0	0.0	0.0	0.0	0.12	1.31
0.0	65.5	0.0	34.2	0.0	0.0	100.0	0.0	0.0	0.0	0.12	0.84
0.0	56.4	0.0	43.6	0.0	0.0	100.0	0.0	0.0	0.0	0.12	1.01
24.9	36.0	0.1	33.0	0.6	11.2	88.4	0.0	0.0	0.4	0.15	1.62
3.5	39.0	0.0	56.4	1.1	0.0	100.0	0.0	0.0	0.0	0.12	2.34
2.9	43.3	0.0	52.9	0.9	0.0	100.0	0.0	0.0	0.0	0.12	2.54
16.9	42.2	0.0	37.1	0.3	0.0	100.0	0.0	0.0	0.0	0.13	1.62
23.2	37.4	0.1	33.5	0.6	9.2	90.4	0.0	0.0	0.3	0.15	1.63
11.0	45.8	0.0	39.2	0.2	0.0	100.0	0.0	0.0	0.0	0.12	1.68
17.9	40.0	0.0	36.6	0.4	5.7	94.1	0.0	0.0	0.2	0.14	1.60
0.0	38.6	0.0	61.4	0.0	0.0	100.0	0.0	0.0	0.0	0.12	0.01
17.6	40.1	0.0	36.9	0.4	5.5	94.3	0.0	0.0	0.2	0.14	1.59

Appendix 1–1. Watershed characteristics for the U.S. Geological Survey streamgage network in the Upper Colorado River Basin.—Continued

Site number	Period of record	Latitude (decimal degrees)	Longitude (decimal degrees)	Drainage area (square miles)	Amount of upstream reservoir storage (acre-feet)	Mean basin elevation (feet)	Mean basin average annual precipitation (inches)	Area covered by developed land (percent)	Area covered by barren land (percent)
09306222	1964–66, 1970 to current year	40.077471	-108 235920	651.72	0	7,297	18.5	0.4	1.7
09306224	1982–89, 2007 to current year	40 158859	-108 343425	1,822.10	7,658	7,889	23.6	0.7	1.0
09306230	1974–77	39 926917	-108.421203	26.03	0	7,325	18.6	0.0	0.9
09306235	1974–89	39 906085	-108 532874	8.65	0	7,760	20.0	0.0	2.7
09306237	1974–82	39 922196	-108 532596	2.73	0	7,590	19.2	0.0	2.6
09306240	1974–85	39.888307	-108 528429	9.18	0	7,876	20.5	0.0	2.6
09306241	1975–82	39 913861	-108.485372	2.51	0	7,077	17.3	0.0	2.7
09306242	1974 to current year	39 920250	-108.472872	31.71	0	7,538	19.2	0.0	3.3
09306244	1975–77	39 933861	-108.427037	37.80	0	7,403	18.7	0.0	2.8
09306246	1975–77	39 967194	-108 388147	5.60	0	6,644	15.9	0.0	0.4
09306248	1975–77	39 981917	-108.453427	39.08	0	7,349	18.6	0.0	3.8
09306250	1975–77	39 980250	-108.408148	49.84	0	7,207	18.0	0.0	3.1
09306255	1972–82, 1988 to current year	40 168581	-108.401205	262.03	0	6,877	16.7	0.0	2.8
09306290	1982 to current year	40 179693	-108 565377	2,531.35	7,658	7,530	21.3	0.5	1.1
09306300	1972–82	40 107198	-108.712882	2.36	0	6,037	12.6	0.4	1.8
09306380	1977–78, 1995	40.087477	-108.776217	424.78	0	6,936	15.9	0.3	1.0
09306395	1977–86	40.013857	-109.080668	3,554.07	7,658	7,201	19.2	0.6	1.4
09306400	1974–76	39 973857	-109 130947	3,576.38	7,658	7,192	19.1	0.6	1.5
09306405	1975–83	39 973302	-109 128447	24.59	0	6,247	11.6	0.1	14.4
09306410	1975–83	39.797749	-109.074554	100.93	0	7,093	15.9	0.7	2.7
09306415	1975–76	39.848303	-109.130668	246.84	0	6,804	14.4	0.4	4.0
09306417	1975–76	39.824969	-109.160946	1.84	0	6,278	11.6	0.0	4.9
09306420	1975–75	39.882747	-109.157336	260.50	0	6,756	14.2	0.4	4.4
09306425	1974–76	39.899969	-109 156225	12.16	0	6,130	11.1	1.0	9.4
09306430	1975–81	39 952190	-109 159281	284.96	0	6,676	13.9	0.4	4.7
09306500	1904–06, 1918, 1923–79, 1985 to current year	39 978856	-109 178727	3,897.42	7,658	7,143	18.6	0.6	1.8
09306600	1975–76	39 954133	-109 258452	3,921.36	7,658	7,132	18.6	0.6	1.8
09306605	1974–76	39 905801	-109 205116	2.57	0	5,891	10.3	0.0	4.0
09306610	1974–76	39 947189	-109 235673	8.37	0	5,592	9.7	0.0	5.6
09306620	1975–76	39 907189	-109 265951	94.98	0	6,119	11.3	0.0	4.2
09306625	1974–83	39 934689	-109 267341	98.36	0	6,092	11.3	0.0	4.4
09306700	1974–77	39 925522	-109 292342	4,027.30	7,658	7,103	18.4	0.5	1.9
09306740	1975–78	39 534140	-109 100388	18.35	0	7,993	19.3	0.0	0.2
09306760	1975–78	39 534971	-109.223170	22.48	0	7,796	18.4	0.0	0.1
09306780	1975–78	39.658580	-109.333452	124.64	0	7,300	16.6	0.0	0.4
09306800	1971–89	39.753301	-109.354841	322.54	0	7,156	15.9	0.0	1.6
09306850	1975–83	39 965521	-109.417071	396.37	0	6,907	14.8	0.0	2.5
09306870	1975–81	39 933577	-109.496796	60.95	0	5,986	10.6	0.0	1.0
09306872	1977–81	39 990521	-109.486797	71.52	0	5,873	10.3	0.0	0.9
09306878	1977–83	40.054131	-109.477353	234.52	0	5,354	9.1	0.8	2.8
09306880	1980–81	40.046631	-109 523744	11.12	0	4,898	7.8	0.0	0.0
09306885	1977–81	40.055798	-109.609025	70.90	0	5,471	9.2	0.0	0.1
09306900	1974–86	40.064964	-109.635692	5,001.30	7,658	6,885	17.0	0.5	2.1
09307000	1948–66	40.072186	-109.728474	35,688.88	5,889,933	7,305	16.4	0.8	1.9
09307200	1976–84	40.082741	-109.873478	156.02	0	5,953	10.0	1.7	1.6
09307290	1978–80	40.031631	-109.752085	300.77	0	5,820	9.6	0.9	1.4
09307300	1975–84	40.029964	-109.757085	300.42	0	5,821	9.6	0.9	1.4
09307500	1951–55, 1958–70, 1975–83	39 566354	-109.587353	299.07	0	7,772	17.4	0.0	0.9
09307800	1975–81	39 525798	-109.734581	89.79	0	8,159	20.0	0.0	0.8
09307900	1975–81	39.876354	-109.704026	273.95	0	7,192	15.3	0.0	4.0
09308000	1948–55, 1975–83	39 938854	-109.648469	899.88	0	7,112	15.0	0.0	3.7
09308500	1950–55, 1957–89	39.798574	-110 565991	31.79	0	8,394	22.2	0.2	3.8

Area covered by deciduous forest (percent)	Area covered by evergreen forest (percent)	Area covered by mixed forest (percent)	Area covered by shrubs young or stunted trees (percent)	Area covered by grass or herbaceous land (percent)	Sedimentary, clastic lithology, Mesozoic (percent)	Sedimentary, clastic (continental) lithology, Tertiary (percent)	Igneous or metamorphic lithologies (percent)	Sedimentary, mixed (continental and marine) lithology (percent)	Sedimentary, carbonate (marine) lithology (percent)	Population density (people per square mile)	Road density (miles per square mile)
15.8	42.0	0.0	37.1	0.3	4.4	95.5	0.0	0.0	0.2	0.13	1.48
24.4	34.4	2.2	25.4	6.0	18.6	53.9	8.7	0.6	18.2	0.34	1.17
9.0	35.6	0.0	54.1	0.0	0.0	100.0	0.0	0.0	0.0	0.12	1.89
18.9	20.3	0.0	57.8	0.3	0.0	100.0	0.0	0.0	0.0	0.12	1.61
6.8	10.8	0.0	79.8	0.0	0.0	100.0	0.0	0.0	0.0	0.12	1.77
20.7	26.7	0.0	49.8	0.1	0.0	100.0	0.0	0.0	0.0	0.12	1.33
0.0	38.6	0.0	58.7	0.0	0.0	100.0	0.0	0.0	0.0	0.12	0.82
11.7	26.7	0.0	58.1	0.1	0.0	100.0	0.0	0.0	0.0	0.12	1.61
9.8	29.3	0.0	57.8	0.1	0.0	100.0	0.0	0.0	0.0	0.12	1.56
0.0	20.7	0.0	78.9	0.0	0.0	100.0	0.0	0.0	0.0	0.12	0.47
6.7	33.3	0.0	55.9	0.2	0.0	100.0	0.0	0.0	0.0	0.11	1.31
5.2	32.6	0.0	58.8	0.2	0.0	100.0	0.0	0.0	0.0	0.11	1.19
3.4	35.6	0.0	57.7	0.1	0.0	100.0	0.0	0.0	0.0	0.09	1.18
18.2	32.6	1.6	37.0	4.4	21.5	56.9	6.2	0.6	14.8	0.26	1.23
0.0	61.5	0.0	35.8	0.0	99.5	0.5	0.0	0.0	0.0	0.04	1.32
8.8	55.9	0.0	32.5	0.7	67.9	32.1	0.0	0.0	0.0	0.19	1.36
14.1	33.8	1.1	42.3	3.2	34.0	50.0	4.4	0.4	11.1	0.29	1.34
14.0	33.7	1.1	42.4	3.2	33.8	50.3	4.4	0.4	11.1	0.29	1.34
0.0	49.3	0.0	36.1	0.0	0.0	100.0	0.0	0.0	0.0	0.04	1.27
6.5	67.5	0.0	21.8	0.2	18.1	81.9	0.0	0.0	0.0	0.08	0.93
3.7	59.0	0.0	32.4	0.1	31.7	68.3	0.0	0.0	0.0	0.06	0.97
0.0	25.2	0.0	69.9	0.0	0.0	100.0	0.0	0.0	0.0	0.04	0.77
3.5	56.5	0.0	34.7	0.1	30.0	70.0	0.0	0.0	0.0	0.06	0.97
0.0	16.8	0.0	72.7	0.0	0.0	100.0	0.0	0.0	0.0	0.04	1.45
3.2	52.7	0.0	38.6	0.1	27.4	72.6	0.0	0.0	0.0	0.05	0.99
13.1	35.1	1.0	42.2	2.9	33.0	52.4	4.0	0.4	10.1	0.27	1.32
13.0	34.9	1.0	42.4	2.9	32.8	52.7	4.0	0.4	10.1	0.27	1.31
0.0	39.6	0.0	56.4	0.0	0.0	100.0	0.0	0.0	0.0	0.04	0.64
0.0	20.9	0.0	73.4	0.0	0.0	100.0	0.0	0.0	0.0	0.04	1.04
0.0	25.6	0.0	70.0	0.0	0.0	100.0	0.0	0.0	0.0	0.04	1.49
0.0	25.0	0.0	70.4	0.0	0.0	100.0	0.0	0.0	0.0	0.04	1.48
12.7	34.6	1.0	43.2	2.8	32.0	53.9	3.9	0.4	9.8	0.26	1.32
35.2	33.0	0.0	25.2	5.1	0.0	100.0	0.0	0.0	0.0	0.11	1.25
14.2	57.2	0.0	27.7	0.1	0.0	100.0	0.0	0.0	0.0	0.05	1.14
3.3	59.9	0.0	35.6	0.0	0.0	100.0	0.0	0.0	0.0	0.04	1.59
6.3	54.5	0.0	36.4	0.3	0.0	100.0	0.0	0.0	0.0	0.05	1.42
5.1	46.7	0.0	44.5	0.3	0.0	100.0	0.0	0.0	0.0	0.04	1.37
0.0	5.6	0.0	93.3	0.1	0.0	100.0	0.0	0.0	0.0	0.04	1.38
0.0	4.7	0.0	94.3	0.1	0.0	100.0	0.0	0.0	0.0	0.04	1.47
0.0	0.2	0.0	95.3	0.7	19.2	80.8	0.0	0.0	0.0	0.07	1.81
0.0	0.0	0.0	100.0	0.0	1.2	98.8	0.0	0.0	0.0	0.04	1.33
0.0	1.0	0.0	98.9	0.0	0.0	100.0	0.0	0.0	0.0	0.04	1.56
10.6	31.8	0.8	49.0	2.3	27.1	61.6	3.2	0.3	7.9	0.22	1.37
6.4	17.5	0.6	62.7	4.8	19.9	64.4	4.6	7.3	3.8	0.57	1.43
0.0	11.6	0.0	75.0	0.3	8.1	91.9	0.0	0.0	0.0	0.26	1.88
0.0	6.8	0.0	84.9	0.3	7.6	92.4	0.0	0.0	0.0	0.26	1.76
0.0	6.8	0.0	84.9	0.3	7.6	92.4	0.0	0.0	0.0	0.26	1.75
5.2	54.2	0.1	36.0	3.0	0.0	100.0	0.0	0.0	0.0	0.05	0.60
21.5	44.5	0.4	26.5	5.4	0.0	100.0	0.0	0.0	0.0	0.00	1.02
7.8	36.1	0.1	49.1	2.1	0.0	100.0	0.0	0.0	0.0	0.03	1.10
4.3	41.6	0.1	47.6	1.7	0.0	100.0	0.0	0.0	0.0	0.04	0.99
27.1	42.7	5.7	20.5	0.0	0.0	100.0	0.0	0.0	0.0	0.39	1.25

Appendix 1–1. Watershed characteristics for the U.S. Geological Survey streamgage network in the Upper Colorado River Basin.—Continued

Site number	Period of record	Latitude (decimal degrees)	Longitude (decimal degrees)	Drainage area (square miles)	Amount of upstream reservoir storage (acre-feet)	Mean basin elevation (feet)	Mean basin average annual precipitation (inches)	Area covered by developed land (percent)	Area covered by barren land (percent)
09309000	1948–55	39.812464	-110 250711	6.67	0	6,793	13.0	4.4	24.0
09309800	1960–69	39.674129	-111 304901	3.62	0	8,962	28.3	3.7	0.0
09310000	1931, 1940–2003	39.715795	-111 300179	12.92	0	8,890	28.0	2.8	0.0
09310500	1931–32, 1938 to current year	39.774406	-111 191006	57.22	0	8,718	26.8	1.2	0.0
09310550	1979–80	39.783017	-111 178228	11.57	0	8,573	25.2	0.0	0.0
09310575	1983–84	39.649963	-111 155727	2.02	0	9,289	28.2	0.3	0.2
09310600	1980–84	39.685241	-111 162672	5.41	0	9,028	26.1	3.0	0.3
09310700	1978–86, 1990 to current year	39.721629	-111 161283	34.52	0	8,935	25.4	1.1	0.2
09311500	1918–21, 1925–31, 1939–69, 1979–80	39.786906	-111 120170	151.60	73,600	8,595	23.7	1.1	0.2
09311700	1962–63	39.827738	-111.009054	1.25	0	8,024	15.9	0.3	3.0
09312000	1942–47	39 933292	-111.067389	22.18	0	8,215	22.9	0.5	0.3
09312500	1938–67	39 922181	-111.057666	54.33	0	8,281	23.1	1.3	0.6
09312600	1967 to current year	39.875793	-111.037388	75.41	0	8,225	22.6	1.3	0.5
09312700	1961–89	39.830516	-110 969330	26.23	0	8,674	21.2	0.5	0.3
09312800	1963–89	39.776907	-110.792381	62.74	0	8,057	20.6	1.5	2.2
09312900	1980–81	39.726907	-110.862106	80.61	0	7,979	20.2	1.6	2.1
09313000	1934–69, 1980–81, 1991–2003	39.718852	-110.865995	455.71	73,600	8,250	21.4	1.2	1.0
09313040	1979–81	39.688575	-110.886551	22.85	0	7,782	17.9	0.3	1.3
09313965	1978–81	39.702464	-110.677935	25.31	0	7,719	18.5	0.7	0.5
09313975	1978–84	39.695242	-110.615157	17.64	0	7,692	17.8	1.7	0.4
09313985	1980–81	39.679409	-110.550435	6.00	0	8,029	18.2	0.5	0.2
09314000	1950–58	39 511077	-110.681267	848.66	73,600	7,585	18.2	1.9	0.9
09314250	1972–86	39.449690	-110.627934	951.34	73,600	7,455	17.6	1.8	0.9
09314280	1972–86	39.421079	-110.646268	190.56	0	5,819	9.8	1.9	1.0
09314340	1978–85	39 555522	-110 380155	39.98	0	8,438	17.7	0.3	0.1
09314374	1978–81	39.457190	-110 359877	12.58	0	7,749	16.3	0.0	0.6
09314500	1945–92, 2000 to current year	39 263858	-110 346546	1,735.19	73,600	6,820	14.7	1.4	1.0
09315000	1894–99, 1905 to current year	38 986083	-110 151248	40,647.49	5,963,533	7,226	16.1	0.7	2.1
09315500	1949–70	38 981362	-110 246808	187.58	0	5,180	9.1	0.6	6.0
09316000	1950–68	38 986083	-110 129858	83.33	0	5,164	10.0	0.6	7.0
09316100	1983–86	38 923306	-109 942350	56.55	0	6,238	12.9	0.0	4.0
09317000	1938–40, 1942–49	39.633297	-111 267398	1.89	0	9,378	28.3	3.2	0.1
09317919	1978–84	39.463299	-111 149060	5.63	0	9,166	29.6	0.0	0.1
09317920	1978–81	39.458577	-111.137115	12.02	0	9,253	27.2	0.6	0.1
09317997	1979 2006	39 385245	-111.088225	178.08	42,400	9,079	26.9	1.2	1.0
09318000	1909–79	39 371356	-111.063780	188.11	42,400	9,019	26.4	1.3	1.2
09318500	1911–17, 1919–21	39 208305	-110 917389	325.43	47,820	7,913	20.1	1.8	1.7
09324000	1954–57	39 283298	-111 267395	147.09	62,460	9,119	26.7	1.4	1.6
09324200	1978–81	39 307188	-111 184615	21.29	0	9,058	26.4	0.6	1.3
09324500	1910–27, 1933–70, 1975–85	39 266633	-111 129892	207.39	62,460	8,902	25.1	1.2	2.3
09325000	1947–58	39 169972	-110 938222	264.31	62,460	8,370	22.1	1.9	2.8
09325100	1965–70	39 149973	-110 909055	683.90	110,280	7,883	19.6	1.8	2.6
09326500	1911–23, 1947 to current year	39 104137	-111 216559	138.43	0	8,863	24.2	1.0	3.6
09327500	1912–14, 1948–58	39 105528	-111.024335	212.74	18,000	7,990	19.4	1.8	4.6
09327550	1976–86	39 119139	-110 989612	218.94	18,000	7,926	19.1	1.8	4.5
09328000	1948–64, 1972–86	39 143584	-110.897944	928.76	128,280	7,832	19.2	1.7	3.0
09328100	1975–86	39.080809	-110.666272	1,319.21	128,280	7,302	16.3	1.3	5.7
09328500	1909–18, 1945 to current year	38.858308	-110.370147	1,668.56	128,280	6,958	14.9	1.1	7.7

Area covered by deciduous forest (percent)	Area covered by evergreen forest (percent)	Area covered by mixed forest (percent)	Area covered by shrubs young or stunted trees (percent)	Area covered by grass or herbaceous land (percent)	Sedimentary, clastic lithology, Mesozoic (percent)	Sedimentary, clastic (continental) lithology, Tertiary (percent)	Igneous or metamorphic lithologies (percent)	Sedimentary, mixed (continental and marine) lithology (percent)	Sedimentary, carbonate (marine) lithology (percent)	Population density (people per square mile)	Road density (miles per square mile)
0.0	14.0	0.0	57.5	0.0	0.0	100.0	0.0	0.0	0.0	0.08	2.05
42.6	2.8	2.3	47.5	0.2	1.3	98.7	0.0	0.0	0.0	0.97	4.80
42.2	6.2	1.5	46.0	0.3	17.7	82.3	0.0	0.0	0.0	0.97	4.27
50.4	16.0	2.6	29.4	0.1	38.4	61.6	0.0	0.0	0.0	0.82	2.47
59.5	18.9	0.6	20.8	0.1	75.4	24.5	0.0	0.0	0.0	0.41	2.20
40.6	43.2	4.1	11.0	0.6	100.0	0.0	0.0	0.0	0.0	0.62	3.68
42.0	37.8	6.7	10.1	0.1	100.0	0.0	0.0	0.0	0.0	0.62	2.59
43.7	38.9	4.8	11.0	0.3	100.0	0.0	0.0	0.0	0.0	0.62	2.59
43.4	19.5	2.4	30.4	0.4	71.1	28.9	0.0	0.0	0.0	0.64	2.69
17.0	1.2	0.2	78.2	0.0	0.0	100.0	0.0	0.0	0.0	0.55	2.24
33.4	40.9	2.0	23.0	0.0	0.0	100.0	0.0	0.0	0.0	0.00	1.47
30.3	35.9	2.7	29.1	0.0	0.0	100.0	0.0	0.0	0.0	0.00	1.31
31.9	29.9	2.7	33.7	0.0	0.0	100.0	0.0	0.0	0.0	0.02	1.83
39.1	22.6	6.5	30.5	0.5	82.5	17.5	0.0	0.0	0.0	0.60	2.90
11.5	22.4	1.5	60.8	0.0	0.0	100.0	0.0	0.0	0.0	0.43	1.26
12.4	29.3	1.4	53.3	0.0	7.1	92.9	0.0	0.0	0.0	0.53	1.30
29.2	24.2	2.2	41.2	0.2	33.7	66.3	0.0	0.0	0.0	0.44	2.30
31.7	53.2	1.9	11.5	0.0	88.4	11.6	0.0	0.0	0.0	0.62	1.54
20.3	24.3	0.5	53.7	0.0	11.1	88.9	0.0	0.0	0.0	0.97	1.54
13.4	38.0	1.2	45.4	0.0	10.1	89.9	0.0	0.0	0.0	0.72	2.09
20.4	52.6	2.2	24.2	0.0	22.7	77.3	0.0	0.0	0.0	0.42	2.35
20.2	25.5	1.5	46.9	0.5	57.1	42.9	0.0	0.0	0.0	3.10	2.35
18.2	24.7	1.4	49.0	0.7	61.2	38.8	0.0	0.0	0.0	2.91	2.32
0.0	2.9	0.0	70.8	7.7	73.9	26.1	0.0	0.0	0.0	0.52	1.82
29.3	61.2	2.2	6.6	0.1	11.2	88.8	0.0	0.0	0.0	0.42	1.14
1.8	92.3	0.2	5.1	0.0	13.9	86.1	0.0	0.0	0.0	0.34	0.92
10.9	21.8	0.9	58.2	2.1	69.2	30.8	0.0	0.0	0.0	1.76	1.94
6.3	19.0	0.6	61.8	4.3	21.4	64.8	4.0	6.4	3.3	0.58	1.41
0.0	0.9	0.0	82.8	9.5	96.5	3.5	0.0	0.0	0.0	0.05	1.12
0.0	16.7	0.0	75.4	0.2	73.6	26.4	0.0	0.0	0.0	0.08	0.68
2.2	41.7	0.0	51.0	0.5	63.1	36.9	0.0	0.0	0.0	0.08	0.53
34.2	15.8	2.0	43.6	1.0	7.2	92.8	0.0	0.0	0.0	1.01	2.66
39.5	45.0	6.4	8.7	0.1	66.3	33.7	0.0	0.0	0.0	0.54	0.17
42.7	31.8	11.7	11.7	1.4	39.0	61.0	0.0	0.0	0.0	0.54	1.77
31.4	44.3	6.1	14.6	0.6	65.7	28.0	0.0	6.3	0.0	0.72	1.52
30.2	44.3	6.0	15.5	0.6	65.1	28.9	0.0	5.9	0.0	0.70	1.60
17.8	31.9	3.6	34.0	2.2	72.0	24.5	0.0	3.4	0.0	1.71	1.72
16.3	38.5	5.8	24.6	10.5	15.4	75.1	0.0	9.4	0.0	0.46	1.49
20.9	50.9	5.6	20.6	0.0	47.3	52.7	0.0	0.0	0.0	0.33	1.72
14.0	39.5	5.0	29.1	7.8	26.9	66.4	0.0	6.7	0.0	0.42	1.40
11.3	33.0	4.1	35.0	6.9	38.0	56.7	0.0	5.2	0.0	1.96	1.67
12.9	29.0	3.3	39.5	4.3	60.3	36.0	0.0	3.7	0.0	1.70	1.69
12.0	35.4	6.9	35.1	5.7	25.6	65.0	0.0	9.4	0.0	0.68	1.29
7.9	25.4	4.5	41.3	7.9	42.4	51.5	0.0	6.1	0.0	0.81	1.56
7.7	24.7	4.4	42.2	8.0	43.6	50.5	0.0	5.9	0.0	0.80	1.55
11.3	27.2	3.5	41.3	5.2	57.5	38.5	0.0	4.1	0.0	1.44	1.65
8.0	19.6	2.4	49.5	8.3	68.1	29.0	0.0	2.9	0.0	1.06	1.51
6.3	16.1	1.9	52.5	10.1	70.5	24.7	0.0	4.3	0.5	0.85	1.45

Site number	Period of record	Latitude (decimal degrees)	Longitude (decimal degrees)	Drainage area (square miles)	Amount of upstream reservoir storage (acre-feet)	Mean basin elevation (feet)	Mean basin average annual precipitation (inches)	Area covered by developed land (percent)	Area covered by barren land (percent)
09329000	1939–45	38.591642	-111.675743	25.57	0	9,685	29.2	2.0	0.3
09329050	1964–98, 2008 to current year	38.627752	-111.647965	24.47	0	10,204	31.8	1.1	1.1
09329500	1949–58	38.485256	-111 576849	183.65	15,582	9,398	23.7	1.5	0.3
09329900	1965–80	38 269426	-111 584072	105.41	0	9,335	22.2	1.3	0.0
09330000	1909–12, 1937–58, 1976 to current year	38 306925	-111 518792	745.76	15,582	8,782	19.2	1.9	0.4
09330210	1969–72	38 272202	-111.092388	115.33	0	7,439	13.0	0.7	12.9
09330230	1967 to current year	38 279147	-111.065720	1,209.20	15,582	8,288	16.7	1.6	5.8
09330410	1983–91	38 121929	-110.759597	7.43	0	9,356	24.0	0.0	0.2
09330500	1909–14, 1949 to current year	38 981919	-111 249336	104.71	0	8,924	23.9	0.4	4.5
09331500	1951–61	38.758308	-111.421568	45.38	0	8,792	19.3	1.2	0.4
09331850	1981–84	38 906361	-111.411847	21.68	0	8,307	17.0	0.8	0.3
09331900	1978–81	38.859142	-111 262115	103.33	0	7,675	14.3	0.7	6.6
09331950	1978–84	38.861365	-111 252670	15.64	0	6,500	9.4	2.5	6.0
09332100	1973–86, 2005–06	38.812198	-111 198780	420.24	0	7,727	15.6	1.2	5.5
09332500	1950–61	38.766642	-111 129335	437.29	0	7,652	15.3	1.2	5.8
09332600	2005–06	38.686944	-110 999528	739.48	0	7,025	12.7	0.9	8.9
09332700	1975–86	38 563034	-110 954327	844.59	0	6,895	12.3	0.8	10.2
09333000	1946–48	38 399982	-110.684038	3,534.39	15,582	6,927	12.5	0.9	14.0
09333500	1948–93, 2001 to current year	38.094153	-110.407364	4,162.20	15,582	6,670	11.9	0.7	16.0
09334000	1951–70	37.898599	-110.449309	136.34	0	5,440	9.4	0.6	31.9
09334500	1951–70	37.798600	-110.376529	277.05	0	6,081	11.3	0.8	6.4
09335500	1950–55	37.766655	-111.684066	92.03	0	8,273	19.1	0.5	0.6
09336000	1950–51	37.762489	-111.738235	35.21	0	8,387	19.8	1.7	1.9
09336500	1952–55	37.766655	-111.684066	104.37	0	7,837	16.4	1.6	2.1
09337000	1950–55, 1957 to current year	37.862487	-111.636011	67.56	0	9,379	25.0	1.0	1.9
09337500	1909–13, 1942–55, 1971 to current year	37.778044	-111 574618	319.06	0	8,115	18.4	1.3	2.1
09338000	1951–55, 1958–72	38.041929	-111.450174	20.38	0	10,689	28.2	1.0	0.1
09338500	1950–55	38.001375	-111 389616	1.62	0	9,252	23.2	2.1	0.0
09338900	2002–07	37.853321	-111 355168	66.09	0	7,635	15.1	1.0	14.1
09339000	1950–55, 2003–07	37.781933	-111 360167	173.52	0	8,329	18.5	1.0	13.1
09339900	1957–96, 1999–2003	37 389730	-106.841151	65.58	0	10,235	41.8	0.0	5.6
09340000	1935–80	37 369451	-106.892262	91.23	0	10,066	40.1	0.0	5.1
09340500	1937–53	37.485562	-106 930321	39.86	0	11,080	50.9	0.0	6.0
09340800	1984–87, 1997–99	37.450285	-106.911709	50.64	0	10,859	48.6	0.0	6.2
09341200	1968–75	37.446396	-106.883931	14.28	0	10,614	46.5	4.0	5.3
09341300	1984–87, 1997–99	37.441952	-106.886986	3.50	0	10,073	39.3	0.7	0.1
09341350	1984–87	37.439174	-106.880042	1.35	0	9,456	31.4	0.5	0.0
09341500	1935–60, 1985–87, 1997–98	37 391951	-106 907263	85.31	0	10,385	43.8	1.0	4.6
09342000	1937–49	37 369451	-106 940318	0.85	0	8,409	24.9	0.0	0.0
09342500	1910–14, 1935 to current year	37 266117	-107.010873	280.65	0	9,739	37.8	0.7	3.8
09343000	1935–71	37 212786	-106.794480	57.92	0	10,023	39.3	0.0	7.6
09343300	1971–98	37 203619	-106.811147	68.68	0	9,922	38.2	0.0	6.8
09343500	1935–52	37 193618	-106 905315	23.24	0	9,252	32.7	0.0	1.4
09344000	1937–95	37.085288	-106.689480	68.93	0	10,261	40.2	0.0	5.1
09344300	1956–70	37.031956	-106.732814	2.32	0	8,456	25.6	0.0	0.0
09344400	1971–98	37.030289	-106.737814	97.48	0	9,862	36.8	0.0	3.9
09345200	1971–96	37.075566	-106.811147	13.29	0	9,605	33.4	0.0	1.4
09345500	1935–52	37.045566	-106.843092	22.99	0	8,946	29.0	1.1	0.8
09346000	1912–96	37.002788	-106.907537	3.55	0	7,763	21.1	0.0	0.0
09346050	2005 to current year	36.961500	-106 977778	230.91	0	8,840	29.1	0.5	1.8
09346400	1961 to current year	37.013617	-107 312267	1,254.09	0	8,393	27.1	0.8	1.6
09347200	1969–75	37.486670	-107 163381	32.06	0	10,428	39.2	0.0	5.9

Area covered by deciduous forest (percent)	Area covered by evergreen forest (percent)	Area covered by mixed forest (percent)	Area covered by shrubs young or stunted trees (percent)	Area covered by grass or herbaceous land (percent)	Sedimentary, clastic lithology, Mesozoic (percent)	Sedimentary, clastic (continental) lithology, Tertiary (percent)	Igneous or metamorphic lithologies (percent)	Sedimentary, mixed (continental and marine) lithology (percent)	Sedimentary, carbonate (marine) lithology (percent)	Population density (people per square mile)	Road density (miles per square mile)
12.1	39.5	9.9	14.6	4.8	13.6	0.0	73.1	13.2	0.0	0.19	2.05
12.2	39.5	11.4	30.0	4.4	0.0	41.2	46.2	12.6	0.0	0.19	1.32
11.6	37.5	9.0	34.5	2.1	7.0	26.9	60.0	6.1	0.0	0.23	1.72
3.5	31.0	9.4	52.1	2.5	2.6	7.5	83.2	6.8	0.0	0.20	1.83
4.7	21.0	4.3	63.7	1.3	8.3	28.7	60.5	2.5	0.0	0.34	1.68
3.8	43.0	2.5	32.2	3.5	70.0	12.4	10.7	4.2	2.7	0.12	1.28
3.6	27.3	3.7	53.0	2.4	31.7	24.0	40.4	3.2	0.8	0.30	1.54
11.5	70.3	3.8	1.0	13.2	39.7	23.4	36.9	0.0	0.0	0.04	1.12
13.4	29.5	6.4	42.0	3.6	38.8	61.2	0.0	0.0	0.0	0.33	1.44
21.5	37.1	4.8	34.2	0.6	44.5	47.5	8.0	0.0	0.0	0.21	1.98
25.9	21.8	0.3	50.8	0.0	85.3	14.7	0.0	0.0	0.0	0.35	3.19
12.7	23.0	0.8	44.0	9.6	82.5	17.5	0.0	0.0	0.0	0.34	1.49
0.0	7.6	0.0	31.2	25.6	65.8	34.2	0.0	0.0	0.0	0.23	2.20
9.6	23.0	2.3	44.1	10.7	63.9	35.2	0.9	0.0	0.0	0.29	1.61
9.3	22.1	2.3	45.2	10.8	64.9	34.3	0.8	0.0	0.0	0.24	1.59
5.5	14.8	1.3	53.6	12.8	74.8	24.2	0.9	0.0	0.0	0.24	1.31
4.8	13.2	1.2	55.5	12.5	77.0	21.2	0.8	0.9	0.0	0.22	1.24
2.8	16.4	1.7	52.2	10.5	66.6	16.6	14.6	1.9	0.3	0.19	1.15
2.3	14.4	1.4	49.3	14.2	71.4	14.3	12.5	1.7	0.2	0.16	1.10
0.5	8.2	0.0	41.9	16.9	94.0	2.5	3.5	0.0	0.0	0.04	0.88
1.6	32.7	0.0	57.2	1.2	33.0	8.5	0.0	58.5	0.0	0.04	1.31
8.4	68.3	7.4	12.5	2.0	49.3	36.3	14.4	0.0	0.0	0.04	0.87
5.0	80.2	7.3	3.7	0.0	26.1	67.4	6.5	0.0	0.0	0.10	1.40
1.9	80.2	3.0	11.1	0.0	38.3	59.5	2.2	0.0	0.0	0.09	1.50
4.9	58.8	20.6	10.5	0.7	15.5	28.5	55.1	0.9	0.0	0.04	1.51
4.2	66.3	7.9	16.0	0.9	41.0	42.0	16.5	0.4	0.0	0.08	1.31
6.8	54.0	7.6	7.8	22.0	10.8	5.1	75.6	8.5	0.0	0.04	1.13
61.3	11.0	10.0	15.1	0.3	43.6	56.4	0.0	0.0	0.0	0.04	2.03
15.0	38.8	1.7	25.8	2.3	69.4	27.7	2.9	0.1	0.0	0.04	0.79
12.1	40.0	3.5	18.4	9.2	50.5	21.9	22.7	4.9	0.0	0.04	1.03
11.4	61.0	6.5	0.5	11.4	10.9	5.0	81.9	2.1	0.0	0.29	0.28
11.5	62.9	7.6	0.5	9.2	7.8	3.6	87.0	1.5	0.0	0.30	0.25
2.8	60.9	0.3	0.0	26.4	0.0	2.8	97.2	0.0	0.0	0.15	0.00
3.2	61.3	2.1	0.0	23.9	0.0	6.2	93.8	0.0	0.0	0.15	0.03
8.6	54.6	0.4	0.1	21.9	0.0	1.3	98.6	0.0	0.0	0.15	1.18
8.3	77.1	2.8	0.0	9.8	0.0	5.7	94.3	0.0	0.0	0.15	0.56
21.3	51.7	25.7	0.0	0.8	0.0	20.0	80.0	0.0	0.0	0.15	0.09
7.8	56.6	5.8	0.3	19.3	0.1	14.6	85.3	0.0	0.0	0.15	0.39
16.6	53.8	25.7	1.8	1.3	73.1	10.4	16.5	0.0	0.0	0.35	0.83
12.1	54.0	9.1	2.9	12.5	18.9	12.1	68.4	0.5	0.0	0.26	0.48
11.9	47.6	11.6	0.4	14.9	0.5	19.5	79.9	0.0	0.0	0.34	0.34
13.6	46.5	13.6	0.6	13.4	1.4	24.7	73.9	0.0	0.0	0.34	0.40
22.7	47.1	16.2	6.5	4.1	35.8	0.0	64.2	0.0	0.0	0.35	1.09
15.2	41.3	7.9	0.3	23.3	1.3	18.4	80.1	0.0	0.0	0.32	0.51
35.3	16.9	13.9	23.6	7.2	100.0	0.0	0.0	0.0	0.0	0.38	1.18
16.7	40.4	8.1	6.7	18.3	13.2	23.3	63.2	0.0	0.0	0.33	0.78
35.6	23.9	15.1	12.5	9.7	10.3	57.5	32.2	0.0	0.0	0.35	0.55
24.2	20.0	9.4	29.4	10.2	34.7	46.7	18.6	0.0	0.0	0.57	0.77
5.1	47.5	0.5	40.4	4.9	100.0	0.0	0.0	0.0	0.0	0.73	2.38
14.3	31.1	5.4	29.3	12.3	48.1	23.1	28.7	0.0	0.0	0.87	1.22
8.8	40.9	4.6	29.3	10.0	58.3	14.5	27.0	0.1	0.0	0.78	1.18
11.7	61.4	5.6	0.3	11.4	2.5	10.9	86.6	0.0	0.0	0.14	0.18

Site number	Period of record	Latitude (decimal degrees)	Longitude (decimal degrees)	Drainage area (square miles)	Amount of upstream reservoir storage (acre-feet)	Mean basin elevation (feet)	Mean basin average annual precipitation (inches)	Area covered by developed land (percent)	Area covered by barren land (percent)
09347205	1978–84	37.452781	-107 176436	34.53	0	10,284	38.2	0.0	5.4
09347500	1936–41, 1946–54	37.428614	-107 193381	82.79	0	10,126	38.1	0.0	5.6
09348500	1936–41, 1946–49	37.461670	-107 198382	44.50	10,084	9,868	33.8	0.0	3.1
09349000	1936–41, 1946–49	37.471669	-107 233105	47.24	0	10,107	29.7	0.0	3.7
09349500	1911–12, 1938–73	37 222226	-107 342826	370.86	10,084	9,434	31.2	0.0	2.1
09349800	1962 to current year	37.088338	-107 397826	654.47	10,084	8,662	27.5	0.6	1.2
09352900	1962 to current year	37.477501	-107 543669	72.55	0	11,347	39.5	0.0	27.8
09353500	1927–86	37 382780	-107 577558	253.42	129,700	10,605	35.8	0.0	11.9
09353800	1999 to current year	37 166111	-107 582500	340.10	129,700	9,936	33.3	0.5	8.8
09354500	1950 to current year	37.009448	-107 599500	518.90	129,700	8,974	28.6	1.1	5.8
09355000	1950 to current year	37.011115	-107 597000	58.38	0	6,938	17.5	1.4	0.0
09355500	1954 to current year	36.801394	-107.698114	3,233.83	1,853,784	8,156	24.5	0.8	1.8
09356500	1907–09, 1910, 1927–55	36.730561	-107.814506	3,502.26	1,853,784	8,024	23.6	0.8	1.7
09356565	1977–81	36.690006	-107.756448	1,700.30	0	6,871	12.8	0.4	0.5
09357000	1909, 1910–11, 1927–31, 1955–63	36.700006	-107 986734	5,352.25	1,853,784	7,599	19.8	0.7	1.3
09357100	1978–81	36.689449	-108.095626	5,463.02	1,853,784	7,564	19.6	0.8	1.3
09357245	1993–94	36.456398	-108.004789	66.72	0	6,613	10.2	0.9	0.3
09357250	1978–81	36.647783	-108 125627	286.85	0	6,263	10.0	2.3	0.2
09357255	1993–94	36.690838	-108 109515	304.46	0	6,232	9.9	2.5	0.2
09357500	1935–82	37.833052	-107 599505	57.43	0	11,932	44.9	0.7	29.4
09358000	1903, 1991–93, 1994 to current year	37.811108	-107.659228	70.32	0	11,820	44.4	0.9	29.8
09358550	1991–93, 1994 to current year	37.819719	-107.663672	20.11	0	11,442	40.3	0.0	19.2
09358900	1968–75	37.851107	-107.725895	11.12	0	11,788	43.1	3.2	16.1
09359000	1935–49	37.797497	-107.695340	5.05	0	11,694	40.6	0.0	29.3
09359010	1991–93, 1994 to current year	37.802774	-107.672839	52.55	0	11,511	40.3	1.5	23.3
09359020	1991 to current year	37.790275	-107.667561	146.09	0	11,622	42.1	1.2	25.6
09359100	1956–61	37.678053	-107.750897	32.44	0	11,094	44.1	1.5	3.8
09359500	1945–56, 2006 to current year	37 570277	-107.780620	349.14	0	11,216	41.0	1.0	20.4
09361000	1911, 1912–14, 1919–28, 1939–80	37.421945	-107.845067	168.54	0	9,592	33.1	0.0	1.1
09361200	1959–65	37 367223	-107.866456	7.32	0	8,821	28.0	0.0	0.0
09361400	1959–65	37 334168	-107 909513	25.65	0	9,406	31.9	0.0	1.1
09361500	1895–1905, 1911 current year	37 279169	-107.880345	701.67	23,254	10,169	35.9	1.2	10.5
09362000	1927–49	37.603886	-107.893680	0.95	0	10,040	35.1	0.0	0.0
09362520	2008 to current year	37 249264	-107.872583	774.52	23,254	9,967	34.8	1.5	9.6
09362550	1995–2002	37 243336	-107.843121	0.44	0	7,205	22.6	0.0	0.0
09362600	1995–97	37 147225	-107.869510	0.61	0	6,569	15.6	1.4	0.0
09362900	1955–63	37.379169	-107.661727	68.64	40,100	10,584	36.9	0.0	2.3
09363000	1899, 1901–03, 1910–12, 1917–24, 1926–60	37 325280	-107.748952	97.26	40,100	10,004	34.7	0.2	1.6
09363050	1967–82	37 295003	-107.791731	107.98	40,100	9,818	33.9	0.3	1.5
09363070	1995–97	37 188615	-107.754229	140.20	40,100	9,225	31.2	1.0	1.1
09363100	1956–63, 1967–83	37 139725	-107.753396	17.96	0	6,760	17.2	2.9	0.0
09363200	1956–63, 1967–83	37.056669	-107.869788	220.77	40,100	8,304	26.1	1.7	0.7
09363500	1933 to current year	37.038058	-107.874232	1,101.96	63,354	9,326	31.4	1.6	6.9
09364010	2002 to current year	36.818337	-108.019791	1,304.70	63,354	8,854	28.6	1.8	5.8
09364500	1904–05, 1912 to current year	36.721392	-108 202018	1,368.99	63,354	8,706	27.7	2.4	5.6
09365000	1904–06, 1912 to current year	36.722780	-108 225630	7,195.26	1,917,138	7,712	20.7	1.2	2.1
09365500	1904–06, 1910, 1917 to current year	37 289722	-108.040628	34.46	0	10,169	38.5	0.5	9.0
09366000	1928–50	37 118888	-108.198689	74.88	0	7,841	21.1	1.2	0.1

Area covered by deciduous forest (percent)	Area covered by evergreen forest (percent)	Area covered by mixed forest (percent)	Area covered by shrubs young or stunted trees (percent)	Area covered by grass or herbaceous land (percent)	Sedimentary, clastic lithology, Mesozoic (percent)	Sedimentary, clastic (continental) lithology, Tertiary (percent)	Igneous or metamorphic lithologies (percent)	Sedimentary, mixed (continental and marine) lithology (percent)	Sedimentary, carbonate (marine) lithology (percent)	Population density (people per square mile)	Road density (miles per square mile)
12.7	60.4	6.4	0.3	10.8	7.8	10.1	81.4	0.7	0.0	0.14	0.29
11.5	53.7	6.0	1.4	14.4	15.9	9.2	73.7	1.2	0.0	0.14	0.61
15.2	48.0	8.7	1.7	16.7	18.5	9.0	67.7	4.8	0.0	0.12	0.46
11.5	56.0	7.1	0.3	14.9	22.8	2.8	68.7	5.7	0.0	0.12	0.38
16.3	54.2	9.7	2.3	11.1	47.7	4.7	35.6	1.9	9.9	0.45	0.65
15.1	51.2	6.7	11.2	10.0	63.9	9.1	20.2	1.1	5.6	1.49	1.19
6.9	36.6	1.3	0.0	25.3	0.0	0.0	61.3	31.7	6.9	1.11	0.05
9.8	49.5	4.7	0.3	20.0	5.0	4.1	61.8	12.1	16.8	0.76	0.28
13.8	47.7	5.1	2.3	15.3	22.1	6.8	46.1	9.2	15.8	1.56	0.86
13.4	40.9	3.7	11.9	11.1	37.2	15.0	30.2	7.1	10.4	2.66	1.47
9.2	29.9	0.5	28.5	2.8	63.0	36.0	0.0	1.0	0.0	1.64	1.89
8.9	43.0	3.7	27.4	8.3	43.2	33.1	19.4	1.4	2.8	1.13	1.46
8.3	41.1	3.5	31.1	8.0	40.0	38.1	17.9	1.3	2.6	1.10	1.56
0.0	15.5	0.0	73.4	10.0	0.0	99.9	0.0	0.0	0.0	0.13	2.66
5.4	31.9	2.3	45.2	9.3	26.2	59.5	11.7	0.9	1.7	0.90	1.98
5.3	31.2	2.2	45.3	9.8	25.6	60.3	11.5	0.8	1.7	1.12	2.02
0.0	3.3	0.0	69.2	26.3	0.0	100.0	0.0	0.0	0.0	0.56	2.49
0.0	1.2	0.0	58.3	30.8	0.0	100.0	0.0	0.0	0.0	0.56	3.28
0.0	1.2	0.0	56.8	31.1	0.0	100.0	0.0	0.0	0.0	0.64	3.29
3.5	12.9	0.0	0.1	50.0	0.0	0.0	96.7	2.9	0.3	0.23	1.08
4.4	14.8	0.0	0.1	46.8	0.0	0.0	97.2	2.6	0.2	0.23	1.07
2.2	41.7	0.2	0.0	35.5	0.0	0.0	100.0	0.0	0.0	0.24	1.05
2.7	25.6	0.1	0.0	50.0	0.0	0.0	100.0	0.0	0.0	0.29	1.95
5.1	18.4	0.1	0.0	40.8	0.0	0.0	64.8	0.0	35.2	0.23	0.00
5.9	29.3	0.3	0.1	36.6	6.7	0.0	81.5	0.0	11.8	0.24	0.96
4.8	24.3	0.1	0.1	41.0	2.4	0.0	91.9	1.2	4.4	0.24	1.11
8.8	44.5	0.7	0.0	38.4	4.0	0.0	27.0	9.0	60.1	0.23	0.40
8.6	34.3	1.1	0.4	31.9	3.2	0.3	70.3	9.5	16.6	0.34	0.71
40.5	45.0	6.4	0.4	6.2	6.0	0.0	0.8	0.0	93.1	0.82	0.55
53.1	18.5	25.0	1.0	0.5	75.5	0.0	0.0	0.0	24.5	0.85	0.95
40.5	39.4	11.3	0.8	6.7	47.2	0.0	0.1	0.0	52.7	1.17	0.76
22.8	38.6	4.0	1.0	18.8	9.8	0.9	38.6	6.4	44.3	3.17	1.08
39.9	56.0	3.9	0.0	0.1	0.0	0.0	0.0	0.0	100.0	1.04	0.00
25.0	37.1	3.9	2.2	17.5	16.8	1.2	35.3	6.0	40.7	4.19	1.26
25.9	51.6	0.0	21.2	0.0	100.0	0.0	0.0	0.0	0.0	12.78	0.44
1.0	16.8	0.0	22.2	0.4	2.4	1.9	0.0	95.7	0.0	1.12	3.64
16.6	41.0	8.6	0.2	28.6	4.1	2.1	45.6	0.0	48.2	0.64	0.90
22.6	43.3	8.2	0.5	20.7	23.4	1.9	32.2	0.0	42.5	1.63	1.05
25.0	42.5	7.7	1.3	18.8	30.4	2.4	29.0	0.0	38.3	2.56	1.29
23.0	40.4	6.0	6.8	14.6	42.5	5.7	22.3	0.0	29.5	3.61	1.76
0.2	28.5	0.0	34.4	2.1	11.4	85.9	0.0	2.7	0.0	3.52	3.32
14.9	38.1	3.8	14.3	9.8	36.6	29.1	14.2	1.4	18.7	3.60	2.31
21.8	37.5	3.5	7.5	14.3	24.1	10.6	27.7	5.3	32.3	3.94	1.60
18.4	35.1	3.0	14.2	15.0	20.5	24.4	23.4	4.5	27.3	4.33	1.88
17.6	33.5	2.8	15.3	15.8	19.6	27.9	22.3	4.3	26.0	7.19	2.09
7.4	30.2	2.2	40.0	12.0	23.3	56.1	13.0	1.4	6.2	2.44	2.11
33.7	22.4	4.6	1.2	24.8	45.0	6.3	21.8	0.0	26.9	1.72	0.95
39.5	27.3	1.0	20.7	3.7	98.3	0.0	0.7	1.0	0.0	2.07	1.06

Site number	Period of record	Latitude (decimal degrees)	Longitude (decimal degrees)	Drainage area (square miles)	Amount of upstream reservoir storage (acre-feet)	Mean basin elevation (feet)	Mean basin average annual precipitation (inches)	Area covered by developed land (percent)	Area covered by barren land (percent)
09366500	1920 to current year	36.999722	-108.188688	308.99	0	7,604	20.1	1.5	1.0
09367000	2002 to current year	36.933334	-108.183965	330.03	0	7,520	19.6	1.5	1.0
09367400	1979–83	36.786113	-108.225909	1.09	0	5,654	9.9	0.0	0.0
09367500	1938–2003, 2005 to current year	36.739724	-108.248131	580.46	0	6,966	16.7	1.2	0.7
09367536	1993–94	36.710558	-108.343689	92.79	0	5,907	9.5	5.2	0.1
09367540	1977–80	36.740279	-108.403135	7,976.87	1,917,138	7,609	20.1	1.4	1.9
09367555	1975–82	36.806389	-108.395636	93.52	0	5,936	11.5	0.0	1.2
09367561	1974–90	36.773334	-108.441193	136.06	0	5,774	10.7	0.1	3.3
09367660	1978–82	35.935296	-107.528106	59.77	0	6,859	10.8	0.2	0.8
09367676	1980–82	36.020016	-107.851726	369.44	0	6,790	10.7	0.2	1.1
09367678	1980–83	36.017516	-107.918395	200.17	0	6,616	10.1	0.6	0.8
09367680	1976–90	36.028627	-107.918395	577.43	0	6,724	10.5	0.3	1.0
09367682	1978–81	36.035016	-107.890894	35.83	0	6,484	9.7	0.0	0.4
09367683	1980–83	36.054182	-107.965063	6.50	0	6,362	9.5	0.0	3.5
09367685	1977–84	36.155013	-107.947008	8.68	0	6,332	9.6	1.0	1.8
09367687	1982–83	35.848630	-108.060621	228.53	0	6,837	10.9	0.7	0.6
09367689	1982–83	35.979182	-108.138124	415.54	0	6,693	10.4	0.6	0.3
09367710	1975–82	36.230843	-108.199794	179.91	0	6,333	9.6	0.3	0.7
09367900	1953–78, 1979–82	35.761128	-108.817310	7.21	0	6,780	11.3	0.0	0.4
09367930	1975–82	36.276953	-108.253963	45.56	0	6,181	9.4	0.0	1.3
09367934	1978–82	36.307229	-108.456747	7.78	0	5,714	8.7	0.0	0.1
09367936	1978–82	36.353062	-108.455080	7.27	0	5,763	8.8	0.0	0.7
09367938	1978–82	36.365839	-108.566472	3,692.07	0	6,415	10.1	0.4	0.9
09367950	1975–94	36.724445	-108.591474	4,390.21	36,550	6,333	9.9	0.5	1.2
09368000	1911, 1927 to current year	36.792220	-108.732311	12,811.99	1,953,688	7,101	16.3	1.1	1.7
09368500	1910–11, 1938–53	37.381663	-108.258137	39.37	0	9,708	33.5	0.1	6.6
09369000	1937–51	37.370274	-108.231469	12.05	0	9,673	33.0	0.1	8.3
09369500	1937–51	37.373885	-108.230636	12.36	0	9,366	30.7	0.0	2.3
09370000	1921, 1931–38	37.357218	-108.254803	71.88	0	9,417	31.5	0.6	5.4
09370600	2006 to current year	37.252496	-108.357862	165.19	15,934	8,260	25.0	1.4	2.4
09370800	1976–79	37.107774	-108.463976	304.56	15,934	7,764	22.1	1.0	1.4
09370820	1979–82	37.099163	-108.466198	322.54	15,934	7,701	21.7	0.9	1.4
09371000	1920–43, 1951 to current year	37.027494	-108.741484	525.78	15,934	7,224	19.0	0.6	1.4
09371002	1986–94	37.200828	-108.697874	26.53	0	6,354	14.4	2.7	0.2
09371010	1977 to current year	37.005553	-109.033992	14,531.04	1,969,622	6,995	15.8	1.0	1.7
09371400	1978–86	37.323883	-108.615094	34.11	0	6,416	14.9	8.2	0.0
09371420	1972–86	37.327216	-108.649262	149.46	0	6,574	15.8	5.1	0.0
09371492	1981–86, 1993 to current year	37.312772	-108.661207	33.72	0	6,340	14.5	3.8	0.0
09371495	1978–81	37.319438	-108.668151	34.12	0	6,334	14.5	3.9	0.0
09371500	1926–29, 1940–45, 1950–54, 1982–93	37.322771	-108.673152	230.42	18,960	6,531	15.5	4.7	0.0
09371520	1993 to current year	37.326660	-108.700653	234.36	18,960	6,524	15.5	4.6	0.0
09371700	1972–83	37.340548	-108.805935	283.05	18,960	6,504	15.3	4.1	0.2
09372000	1951 to current year	37.324160	-109.015666	346.38	18,960	6,408	14.9	3.5	0.4
09372200	1981–82	37.216664	-109.184001	717.17	18,960	6,182	14.1	2.5	0.7
09372400	2001–08	37.876660	-109.445954	2.83	0	10,013	33.2	0.0	0.5
09376800	2001–08	37.888882	-109.466510	2.53	0	9,953	33.3	0.0	0.0
09376900	1966–72	37.919437	-109.434842	4.91	0	9,184	29.4	0.0	0.0
09378100	1980–85	37.873049	-109.366507	5.64	0	8,312	23.7	0.1	0.0
09378170	1985 to current year	37.846661	-109.369562	8.45	0	8,614	26.0	0.0	0.1
09378200	1979–92	37.860550	-109.342339	21.69	0	8,168	23.2	1.1	0.1
09378490	2001–05	37.811380	-109.024276	97.24	0	7,038	15.1	3.1	0.0
09378600	1985–93	37.299998	-109.300671	1,155.59	0	6,358	14.2	2.1	0.2

Area covered by deciduous forest (percent)	Area covered by evergreen forest (percent)	Area covered by mixed forest (percent)	Area covered by shrubs young or stunted trees (percent)	Area covered by grass or herbaceous land (percent)	Sedimentary, clastic lithology, Mesozoic (percent)	Sedimentary, clastic (continental) lithology, Tertiary (percent)	Igneous or metamorphic lithologies (percent)	Sedimentary, mixed (continental and marine) lithology (percent)	Sedimentary, carbonate (marine) lithology (percent)	Population density (people per square mile)	Road density (miles per square mile)
19.2	23.9	0.8	34.0	8.1	77.5	1.9	2.6	15.0	3.0	1.71	1.50
18.0	23.8	0.8	35.5	8.2	77.7	3.0	2.4	14.0	2.8	1.69	1.54
0.0	0.0	0.0	73.5	26.4	66.3	33.7	0.0	0.0	0.0	1.89	2.96
10.5	21.0	0.4	46.2	12.2	66.2	22.6	1.4	8.2	1.6	1.95	1.95
0.0	0.0	0.0	48.7	14.3	54.7	45.3	0.0	0.0	0.0	1.45	2.65
7.4	28.8	2.0	40.7	12.2	27.1	53.4	11.8	1.9	5.7	2.75	2.14
0.0	1.5	0.0	87.2	10.1	98.9	1.1	0.0	0.0	0.0	0.27	2.48
0.0	1.0	0.0	83.8	11.2	99.3	0.7	0.0	0.0	0.0	0.50	2.72
0.0	6.8	0.0	33.9	58.3	82.8	16.9	0.0	0.0	0.0	0.62	1.29
0.0	7.4	0.0	31.1	60.2	59.1	40.8	0.0	0.0	0.0	0.60	2.18
0.0	1.2	0.0	40.0	57.4	100.0	0.0	0.0	0.0	0.0	0.16	1.50
0.0	5.3	0.0	34.3	59.1	73.8	26.1	0.0	0.0	0.0	0.44	1.92
0.0	0.2	0.0	45.6	53.9	98.5	1.5	0.0	0.0	0.0	0.35	0.87
0.0	2.3	0.0	63.8	30.4	100.0	0.0	0.0	0.0	0.0	0.35	0.78
0.0	0.0	0.0	83.5	13.7	100.0	0.0	0.0	0.0	0.0	0.35	0.30
0.0	5.1	0.0	72.5	21.1	99.9	0.0	0.0	0.0	0.0	1.26	1.65
0.0	2.9	0.0	61.5	34.6	100.0	0.0	0.0	0.0	0.0	0.78	1.44
0.0	0.9	0.0	79.6	18.5	60.9	39.1	0.0	0.0	0.0	0.53	1.09
0.0	29.7	0.0	28.5	41.4	100.0	0.0	0.0	0.0	0.0	1.24	2.54
0.0	0.0	0.0	72.0	26.7	69.8	30.2	0.0	0.0	0.0	0.58	1.50
0.0	0.0	0.0	86.0	13.9	100.0	0.0	0.0	0.0	0.0	0.12	1.29
0.0	0.0	0.0	80.2	19.1	100.0	0.0	0.0	0.0	0.0	0.12	2.72
0.0	7.0	0.0	55.6	35.9	82.1	17.8	0.1	0.0	0.0	0.81	1.76
0.0	7.4	0.0	57.3	33.1	83.2	16.6	0.1	0.0	0.0	0.87	1.79
4.6	20.6	1.3	47.7	19.4	48.7	39.1	7.4	1.2	3.6	2.10	2.05
41.3	32.8	7.3	1.2	9.6	93.3	0.0	6.7	0.0	0.0	0.73	1.16
49.3	20.6	9.2	0.6	10.8	93.0	0.0	0.1	0.7	6.2	0.74	0.68
59.5	18.4	7.8	0.6	9.5	89.9	0.0	10.1	0.0	0.0	0.73	1.32
44.6	27.9	6.9	2.0	8.9	91.6	0.0	5.4	1.9	1.0	0.76	1.28
26.7	33.8	3.1	15.3	6.1	91.1	0.2	2.4	5.8	0.5	1.09	1.96
18.4	35.1	1.8	23.0	12.4	95.1	0.1	1.3	3.2	0.2	1.08	1.62
17.4	35.1	1.7	25.2	11.7	95.4	0.1	1.3	3.0	0.2	1.03	1.58
11.4	30.7	1.1	41.9	8.7	97.1	0.1	0.8	1.8	0.1	0.76	1.22
5.0	19.5	0.2	55.3	1.6	59.2	35.1	5.8	0.0	0.0	2.84	1.83
4.5	20.1	1.2	49.2	18.7	54.2	34.8	6.7	1.1	3.2	1.98	2.03
0.5	13.9	0.0	35.7	1.4	100.0	0.0	0.0	0.0	0.0	21.24	5.14
1.5	21.0	0.0	47.7	2.1	95.5	4.5	0.0	0.0	0.0	13.98	3.40
3.4	25.0	0.0	50.6	1.7	69.4	28.9	1.8	0.0	0.0	2.63	2.20
3.4	24.8	0.0	50.8	1.7	69.7	28.5	1.8	0.0	0.0	2.65	2.22
1.5	20.6	0.0	45.7	1.8	92.6	7.2	0.3	0.0	0.0	9.86	3.12
1.5	20.7	0.0	46.0	1.8	92.7	7.0	0.3	0.0	0.0	9.73	3.09
1.4	27.2	0.0	42.6	1.5	93.1	6.1	0.7	0.0	0.0	8.25	2.84
1.8	28.3	0.1	45.1	1.3	93.0	5.4	1.6	0.0	0.0	6.90	2.44
1.0	26.3	0.0	48.2	3.1	95.9	3.1	1.0	0.0	0.0	3.72	2.00
25.4	39.0	14.5	0.2	20.4	43.1	0.0	56.9	0.0	0.0	0.08	1.60
16.2	32.7	34.9	0.2	16.0	47.4	0.0	52.6	0.0	0.0	0.08	0.13
41.4	23.9	19.6	4.5	10.5	64.1	5.8	30.1	0.0	0.0	3.12	0.97
50.5	22.7	7.3	14.5	4.3	88.3	0.0	11.7	0.0	0.0	3.29	1.96
55.3	19.6	3.4	7.7	13.1	71.6	0.7	27.7	0.0	0.0	0.08	1.25
40.6	29.1	3.3	16.5	6.4	83.7	2.4	14.0	0.0	0.0	1.08	2.04
1.7	18.4	0.0	52.6	0.1	99.9	0.1	0.0	0.0	0.0	0.24	1.77
2.3	27.8	0.2	48.4	0.9	96.5	2.7	0.8	0.0	0.0	0.63	1.73

Appendix 1–1. Watershed characteristics for the U.S. Geological Survey streamgage network in the Upper Colorado River Basin.—Continued

Site number	Period of record	Latitude (decimal degrees)	Longitude (decimal degrees)	Drainage area (square miles)	Amount of upstream reservoir storage (acre-feet)	Mean basin elevation (feet)	Mean basin average annual precipitation (inches)	Area covered by developed land (percent)	Area covered by barren land (percent)
09378630	1965 to current year	37.755550	-109.476511	4.02	0	8,615	24.6	0.0	0.0
09378650	1975–93	37.680830	-109.462621	37.39	0	7,912	21.5	0.6	0.1
09378700	1965–87	37 560554	-109 578735	204.39	0	6,805	15.2	0.4	2.9
09379000	1959–68	37 266112	-109.675678	278.27	0	5,859	10.5	0.4	8.5
09379200	1964 to current year	36 943891	-109.710668	3,655.05	12,500	6,244	10.1	0.7	6.5
09379500	1914 to current year	37 146946	-109.864844	23,020.45	2,001,082	6,620	14.0	1.0	3.4
09379505	1928–84	37 147224	-109.865122	23,020.58	2,001,082	6,620	14.0	1.0	3.4
09379910	1989–2004	36 921655	-111.483491	107,840.88	37,685,964	7,069	15.7	0.9	5.3
09380000	1895 to current year	36.864710	-111 588216	108,094 34	37,685,964	7,065	15.7	0.9	5.3
385106 106571000	2006 to current year	38.851659	-106 953377	73.05	0	10,272	32.7	2.3	7.6
392547 106023400	2004 to current year	39.429722	-106.042778	21.06	0	11,810	31.4	1.6	30.1
394220 106431500	2006 to current year	39.705000	-106.725833	606.41	0	9,950	24.8	2.5	7.4
400016 105490800	1997–99	40.004431	-105.819454	14.17	0	9,980	31.0	0.0	2.8

Area covered by deciduous forest (percent)	Area covered by evergreen forest (percent)	Area covered by mixed forest (percent)	Area covered by shrubs young or stunted trees (percent)	Area covered by grass or herbaceous land (percent)	Sedimentary, clastic lithology, Mesozoic (percent)	Sedimentary, clastic (continental) lithology, Tertiary (percent)	Igneous or metamorphic lithologies (percent)	Sedimentary, mixed (continental and marine) lithology (percent)	Sedimentary, carbonate (marine) lithology (percent)	Population density (people per square mile)	Road density (miles per square mile)
48.7	38.5	2.1	0.4	10.1	80.7	0.0	19.3	0.0	0.0	0.06	1.62
38.0	43.1	1.7	7.4	6.0	90.5	0.7	8.3	0.5	0.0	0.56	1.85
11.0	55.6	0.1	26.0	1.0	81.6	6.6	0.6	10.6	0.5	0.04	1.49
2.4	29.2	0.0	56.6	1.5	9.1	1.6	0.0	88.0	1.2	0.04	0.94
0.1	18.9	0.0	57.9	15.3	92.1	5.2	0.4	2.3	0.0	1.33	2.28
3.2	19.3	0.7	51.6	16.6	65.8	23.6	4.5	3.7	2.4	1.68	2.04
3.2	19.3	0.7	51.6	16.6	65.8	23.6	4.5	3.7	2.4	1.68	2.04
7.4	21.6	0.8	50.2	9.4	46.4	35.2	9.0	5.5	3.9	1.11	1.47
7.4	21.5	0.8	50.2	9.4	46.5	35.2	8.9	5.5	3.9	1.11	1.47
18.2	34.9	0.4	1.5	27.9	63.8	8.6	27.6	0.0	0.0	4.45	2.29
1.2	33.1	0.0	0.0	24.6	0.0	0.0	65.9	0.0	34.0	3.73	2.29
16.6	40.3	1.9	13.7	14.3	11.5	5.3	38.9	1.8	42.4	5.33	1.78
0.2	85.6	0.0	0.0	3.9	0.0	0.0	100.0	0.0	0.0	1.51	1.87

www.ingramcontent.com/pod-product-compliance
Lightning Source LLC
Chambersburg PA
CBHW081548170526
45166CB00009B/2623